Careers in Focus

Engineering

Ferguson Publishing Company
Chicago, Illinois

Copyright © 1999 Ferguson Publishing Company
ISBN 0-89434-282-7

Library of Congress Cataloging-in-Publication Data

Careers in focus.   Engineering.
       p.   cm.
    Summary: Defines the top twenty careers in engineering in terms of the
nature of the work, educational or training requirements, ways to get started,
advancement possibilities, salary figures, employment outlook, and sources of
more information.
    ISBN 0-89434-282-7
    1. Engineering—Vocational guidance. [1. Engineering—Vocational guid-
ance. 2. Vocational guidance.] I. Title: Engineering.
TA157.C283 1999
620'.0023—dc21   99-36146
                                                           CIP

Printed in the United States of America

Cover photo courtesy Tony Stone Images

Published and distributed by
Ferguson Publishing Company
200 West Jackson Boulevard, 7th Floor
Chicago, Illinois 60606
312-692-1000

W-8

Table of Contents

Introduction

— There are 1.4 million engineers in the United States. *Engineer* is a broad term that applies to people who work in several different fields. In addition to sharing the job title of engineer, these professionals share a strong math and science background and are relied on for their ability to develop solutions to practical problems. Beyond these common factors, the jobs engineers do vary greatly, depending on their specialty. All engineers are problem solvers and, to some extent, inventors. An aerospace engineer, for example, may spend much of his or her career working on a way to prevent ice from forming on commercial aircraft. In the course of working on a solution to this problem, the engineer may work with chemical engineers to invent a chemical solution that can be applied to planes. Or the aerospace engineer may spend years testing engine components to find parts that work together to resist the effects of ice. However they arrive at solutions to such problems, all engineers must keep practical matters in mind, such as whether the cost of the new process will be more expensive than the original problem.

Regardless of their specialty, engineers are all involved with highly technical work. They have to have a thorough knowledge of how things work—from electronics to the human body—in order to be able to come up with better ways of doing things. Civil engineers who design bridges and highways, for example, have to be comfortable with using the laws of physics to determine how much weight a structure will hold and how best to distribute that weight.

As a rule, engineers all have at least a four-year degree that provides them with a clear understanding of how math and science applies to the everyday world. Most engineering degrees are granted in the broad areas of electrical, mechanical, or civil engineering. Graduates of these programs may then choose to further specialize in their area of interest by undergoing more college training or by learning the basics of a specialty on the job. For example, a mechanical engineer who wants to work with nuclear reactors may study nuclear science beyond the undergraduate level or may take an entry-level engineering job at a nuclear plant before working up to becoming an actual nuclear engineer.

Because so much of the work they do affects the safety of the public, engineers usually must be licensed in the states in which they work. All 50 states require registration for engineers whose work may affect life, health, or property, or who offer their services to the public. These guidelines cover the great majority of engineers. Registration generally requires an undergraduate degree from a program accredited by the Accreditation Board of Engineering and Technology, four years of relevant professional experience, and successful completion of a state examination.

The work of engineers is essential to maintain and improve our quality of life, and they are paid well for it. Engineering has among the highest starting salaries of any career. In 1997, new engineers with a bachelor's degree averaged $38,500, according to the National Association of Colleges and Employers. Those with a master's degree and no experience earned an average of $45,400 a year. Engineers with several years of experience and education can earn six figures in a year.

Each article in this book discusses a particular engineering occupation in detail. The information comes from Ferguson's *Encyclopedia of Careers and Vocational Guidance.* The History section describes the history of the particular job as it relates to the overall development of its industry or field. The Job describes the primary and secondary duties of the job. Requirements discusses high school and postsecondary education and training requirements, any certification or licensing necessary, and any other personal requirements for success in the job. Exploring offers suggestions on how to gain some experience in or knowledge of the particular job before making a firm educational and financial commitment. The focus is on what can be done while still in high school (or in the early years of college) to gain a better understanding of the job. The Employers section gives an overview of typical places of employment for the job. Starting Out discusses the best ways to land that first job, be it through the college placement office, newspaper ads, or personal contact. The Advancement section describes what kind of career path to expect from the job and how to get there. Earnings lists salary ranges and describes the typical fringe benefits. The Work Environment section describes the typical surroundings and conditions of employment—whether indoors or outdoors, noisy or quiet, social or independent, and so on. Also discussed are typical hours worked, any seasonal fluctuations, and the stresses and strains of the job. The Outlook section summarizes the job in terms of the general economy and industry projections. For the most part, Outlook information is obtained from the Bureau of Labor Statistics and is supplemented by information taken from professional associations. Job growth terms follow those used in the *Occupational Outlook Handbook:* Growth described as "much faster than the average" means an increase of 36 percent or more. Growth described as "faster than the average" means an increase of 21 to 35 percent. Growth described as "about as fast as the average" means an increase of 10 to 20 percent. Growth described as "little change or more slowly than the average" means an increase of 0 to 9 percent. "Decline" means a decrease of 1 percent or more.

Each article ends with For More Information, which lists organizations that can provide career information on training, education, internships, scholarships, and job placement.

Aerospace Engineers

Mathematics
Physics
— School Subjects

Mechanical/manipulative
Technical/scientific
— Personal Skills

Primarily indoors
Primarily one location
— Work Environment

Bachelor's degree
— Minimum Education Level

$37,957 to $57,000 to $120,000+
— Salary Range

Required by certain states
— Certification or Licensing

Decline
— Outlook

Overview

Aerospace engineering encompasses the fields of aeronautical (aircraft) and astronautical (spacecraft) engineering. *Aerospace engineers* work in teams to design, build, and test machines that fly within the earth's atmosphere and beyond. Although aerospace science is a very specialized discipline, it is also considered one of the most diverse. This field of engineering draws from such subjects as physics, mathematics, earth science, aerodynamics, and biology. Some aerospace engineers specialize in designing one complete machine, perhaps a commercial aircraft, whereas others focus on separate components such as for missile guidance systems. There were about 50,000 aerospace engineers in the United States in 1996, according to the Bureau of Labor Statistics.

History

The roots of aerospace engineering can be traced as far back as when people first dreamed of being able to fly. Thousands of years ago, the Chinese developed kites and later experimented with gun powder as a source of propulsion. In the 15th century, Renaissance artist Leonardo da Vinci created drawings of two devices that were designed to fly. One, the ornithopter, was supposed to fly the way birds do, by flapping its wings; the other was designed as a rotating screw, closer in form to today's helicopter.

In 1783, Joseph and Jacques Montgolfier of France designed the first hot-air balloon that could be used for manned flight. In 1799 an English baron, Sir George Cayley, designed an aircraft that was one of the first not to be considered "lighter than air," as balloons were. He developed a fixed-wing structure that led to his creation of the first glider in 1849. Much experimentation was performed in gliders and the science of aerodynamics through the late 1800s. In 1903, the first mechanically powered and controlled flight was completed in a craft designed by Orville and Wilbur Wright. The big boost in airplane development occurred during World War I. In the early years of the war, aeronautical engineering comprised a variety of engineering skills applied toward the development of flying machines. Civil engineering principles were used in structural design, while early airplane engines were devised by automobile engineers. Aerodynamic design itself was primarily empirical, with many answers coming from liquid-flow concepts established in marine engineering.

The evolution of the airplane continued during both world wars, with steady technological developments in materials science, propulsion, avionics, and stability and control. Airplanes became larger and faster. Airplanes are commonplace today, but commercial flight became a frequent mode of transportation only as recently as the 1960s and 1970s.

Robert Goddard developed and flew the first liquid-propelled rocket in 1926. The technology behind liquid propulsion continued to evolve, and in 1938 the first U.S. liquid rocket engine was tested. More sophisticated rockets were eventually created to enable aircraft to be launched into space. The world's first artificial satellite, *Sputnik I,* was launched by the Soviets in 1957. In 1961, President John F. Kennedy urged the United States to be the first country to put a man on the moon; on July 20, 1969, astronauts Neil Armstrong and Edwin Aldrin, Jr., accomplished that goal.

Today, aerospace engineers design spacecraft that explore beyond the earth's atmosphere, such as space shuttles and rockets. They create missiles and military craft of many types, such as fighters, bombers, observers, and transports. Today's engineers go way beyond the dreams of merely learning to fly. For example, the International Space Station, the newest space venture, is an effort between the United States and 15 other countries. It will take a total of 43 missions, beginning in fall 1998, to fully assemble the station. Space professionals on the ground, including aerospace engineers, will play a vital role in developing equipment that will be used on the station.

The Job

Although the creation of aircraft and spacecraft involve professionals from many branches of engineering (e.g., materials, electrical, and mechanical), aerospace engineers in particular are responsible for the total design of the craft, including its shape, performance, propulsion, and guidance control system. In the field of aerospace engineering, professional responsibilities vary widely depending on the specific job description. Aeronautical engineers work specifically with aircraft systems, and astronautical engineers specialize in spacecraft systems.

Throughout their education and training, aerospace engineers thoroughly learn the complexities involved in how materials and structures perform under tremendous stress. In general, they are called upon to apply their knowledge of the following subjects: propulsion, aerodynamics, thermodynamics, fluid mechanics, flight mechanics, and structural analysis. Less technically scientific issues must also often be dealt with, such as cost analysis, reliability studies, maintainability, operations research, marketing, and management.

There are many professional titles given to certain aerospace engineers. *Analytical engineers* use engineering and mathematical theory to solve questions that arise during the design phase. *Stress analysts* determine how the weight and loads of structures behave under a variety of conditions. This analysis is performed with computers and complex formulas.

Computational fluid dynamic (CFD) engineers use sophisticated high-speed computers to develop models used in the study of fluid dynamics. Using simulated systems, they determine how elements flow around objects; simulation saves time and money and eliminates risks involved with actual testing. As computers become more complex, so will the tasks of the CFD engineer.

Design aerospace engineers draw from the expertise of many other specialists. They devise the overall structure of components and entire crafts, meeting the specifications developed by those more specialized in aerodynamics, astrodynamics, and structural engineering. Design engineers use computer-aided design (CAD) programs for many of their tasks. *Manufacturing aerospace engineers* develop the plans for producing the complex components that make up aircraft and spacecraft. They work with the designers to ensure that the plans are economically feasible and will produce efficient, effective components.

Materials aerospace engineers determine the suitability of the various materials that are used to produce aerospace vehicles. Aircraft and spacecraft require the appropriate tensile strength, density, and rigidity for the particular environments they are subjected to. Determining how materials such as steel, glass, and even chemical compounds react to temperature and stress is an important part of the materials engineer's responsibilities.

Quality control is a task that aerospace engineers perform throughout the development, design, and manufacturing processes. The finished product must be evaluated for its reliability, vulnerability, and how it is to be maintained and supported.

Marketing and sales aerospace engineers work with customers, usually industrial corporations and the government, informing them of product performance. They act as a liaison between the technical engineers and the clients to help ensure that the products delivered are performing as planned. Sales engineers also need to anticipate the needs of the customer, as far ahead as possible, to inform their companies of potential marketing opportunities. They also keep abreast of their competitors and need to understand how to structure contracts effectively.

Requirements

High School

High school students interested in aerospace careers should follow a college-preparatory program. Doing well in mathematics and science classes is vital to students who want to pursue a career in any type of engineering field. The American Society for Engineering Education advises students interested in engineering to take calculus and trigonometry in high school, as well as laboratory science classes. Such courses provide the skills you'll need for problem solving, an essential skill in any type of engineering.

Postsecondary Training

Aerospace engineers need a bachelor's degree to enter the field. More advanced degrees are optional, depending on the type of position you're seeking. College admissions officers will expect students entering engineering programs to have high scores in the math and science sections of entrance examinations such as the ACT and SAT.

While a major in aerospace engineering is the norm, other majors are sometimes acceptable. For example, the National Aeronautics and Space Administration (NASA) recommends a degree in any of a variety of disciplines, including biomedical engineering, ceramics engineering, chemistry, industrial engineering, materials science, metallurgy, optical engineering, and oceanography. You should make sure the college you choose has an accredited engineering program. The Accreditation Board for Engineering and Technology must meet minimum education standards in these fields. Graduation from an ABET-accredited school is a requirement for becoming licensed in many states, so it is important that you select an accredited school. Currently, more than 300 colleges and universities offer ABET-accredited engineering programs. Your high school counselor can help you find a list of accredited schools.

Many aerospace engineers complete master's degree and even doctoral work before entering this field. Advanced degrees can significantly increase an engineer's earnings. Students continuing on to graduate school are concerned more with research and further specialization, with a thesis required for a master's degree and a dissertation required for a doctorate. About one-third of all aerospace engineers go on to graduate school to get a master's degree.

Certification or Licensing

Most states require engineers to be licensed. There are two levels of licensing for engineers. Licensed Professional Engineers are engineers who have graduated from an accredited engineering curriculum, have four years of engineering experience, and have passed a written exam. Engineering graduates need not wait until they have four years experience, however, to become licensed, according to the National Society of Professional Engineers. Most states provide a prelicensure certificate for those without four years' experience. This certification is known as Engineer-in-Training. Requirements vary by state, but this prelicensing also usually requires graduation from an accredited institution and successful completion of an engineering exam administered by the state.

Exploring

Students who like to work on model airplanes and rockets may be good candidates for an aerospace engineering career. Working on special research assignments supervised by science and math teachers is helpful, as is working on cars and boats, which provides a good opportunity to discover more about aerodynamics. A part-time job with a local manufacturer can give students some exposure to product engineering and development.

Exciting opportunities are often available at summer camps and academic programs throughout the country. For instance, the International Aerospace Camp (see address listed at the end of this article) presents a 10-day session focusing on aerospace study and career exploration. Instruction in model rocketry and flight are also offered. In addition, the camp provides jobs for qualified college students. Admission to the camp is competitive, though; the camp usually consists of two 10-day programs for 32 students each. It is also a good idea to join a science club. For example, the Junior Engineering Technical Society provides member students with an opportunity to enter academic competitions, explore career opportunities, and design model structures. The *JETS Report* is one type of journal that students should read; it offers articles on industry and student news and club activities. JETS also administers a high-school level competition for students interested in engineering. Information is available at the JETS Web site, listed at the end of this article.

Starting Out

Many students begin their careers while completing their studies; many work-study jobs lead to full-time careers. Most aerospace manufacturers actively recruit engineering students, conducting on-campus interviews and other activities to locate the best candidates. Students preparing to graduate can also send out resumes to companies active in the aerospace industry and arrange interviews. Many colleges and universities also staff job placement centers, which are often good places to find leads for new job openings.

Students can also apply directly to agencies of the federal government concerned with aerospace development and implementation. Applications can be made through the Office of Personnel Management or through an agency's own hiring department.

Professional associations, such as the National Society of Professional Engineers and the American Institute of Aeronautics and Astronautics, offer job placement services, including career advice, job listings, and training. Their Web addresses are listed at the end of this article.

Advancement

As in most engineering fields, in the various divisions of aerospace engineering there tends to be a hierarchy of workers. This is true in research, design and development, production, and teaching. In an entry-level job, one is considered simply an engineer, perhaps a junior engineer. After a certain amount of experience is gained, depending on the position, one moves on to work as a project engineer, supervising others. Then, as a managing engineer, one has further responsibilities over a number of project engineers and their teams. At the top of the hierarchy is the position of chief engineer, which involves authority over managing engineers and additional decision-making responsibilities.

As engineers move up the career ladder, the type of responsibilities they have tend to change. Junior engineers are highly involved in technical matters and scientific problem solving. As managers and chiefs, engineers have the responsibilities of supervising, cost analyzing, and relating with clients.

All engineers must continue to learn and study technological progress throughout their careers. It is important to keep abreast of engineering advancements and trends by reading industry journals and taking courses. Such courses are offered by professional associations or colleges. In aerospace engineering especially, changes occur rapidly, and those who seek promotions must be prepared. Those who are employed by colleges and universities must continue teaching and conducting research if they want to have tenured (more guaranteed) faculty positions.

Earnings

Starting salaries for aerospace engineers with bachelor's degrees averaged $37,957 per year in 1996, according to the Bureau of Labor Statistics. With experience, the average salaries for all aerospace engineers is about $57,700 per year. Many employers of engineers use a table ranging from Engineer I to Engineer VIII. The categories are based on level of education, duties and job experience. According to the National Society for Professional Engineers, the average starting salary for all new engineering graduates (Engineer I) was $41,487 in 1998. Salary recommendations for mid-level engineers (Engineer IV and V) ranged from $62,230 to $76,750. Salaries for the highest level of engineer (Engineer VIII) were $124,461 or higher.

Engineers employed by the federal government generally earn less than those in private industry, with an average salary for experienced engineers of around $54,000 per year. Federal employees, however, enjoy greater job security and often more generous vacation and retirement benefits.

All engineers can expect to receive vacation and sick pay, paid holidays, health insurance, life insurance, and retirement programs.

Work Environment

Aerospace engineers work in various settings depending on their job description. Many aircraft-related engineering jobs are found in Texas, Washington, and California, where large aerospace companies are located. Those involved in research and design usually work at desks in well-lit, air-conditioned offices. They spend much time at computers and drawing boards. Engineers involved with the testing of components and structures often work outside at test sites or at locations where controlled testing conditions can be created; these places can be considered large laboratories.

In the manufacturing area of the aerospace industry, engineers often work on the factory floor itself, assembling components and making sure that they conform to design specifications. This job requires much walking around large production facilities, like aircraft factories or spacecraft assembly plants.

Engineers are sometimes required to travel to other locations to consult with companies that make materials and other needed components. Many also go to remote test sites to observe and participate in flight testing.

Aerospace engineers are also employed with the Federal Aviation Administration (FAA) and commercial airline companies. Here they may perform a variety of duties, including performance analysis and crash investigations. Companies that are involved with satellite communications need the expertise of aerospace engineers to better interpret the many aspects of the space environment and the problems involved with getting a satellite launched into space.

Outlook

Employment in the aerospace industry has been down for many years. Between 1989 and 1995, over 500,000 aerospace jobs were lost. The aerospace industry has gone through difficult times since the late 1980s, and more job losses are predicted for the immediate future. The collapse of the Soviet Union, shrinking defense and space program budgets, the recession of the early 1990s, and the continuing wave of corporate downsizing all combined to cut severely into the aerospace industry. Pressure to balance the federal budget, for example, has greatly reduced spending for aerospace. Nevertheless, the aerospace industry remains vital to the health of the national economy. Advances in aircraft design, which have produced quieter and more fuel-efficient aircraft, will make them attractive to airlines seeking to replace their increasingly aging fleets.

Despite cutbacks in the space program, the development of new space technology and increasing commercial uses for that technology will continue to demand qualified engineers. Facing reduced demand in the United States, aerospace companies are increasing their sales overseas, and depending on the world economy and foreign demand, this new market could create a demand for new workers in the industry.

Employment opportunities within aerospace will remain intensely competitive, however. Manufacturers and government agencies will seek only the top students to fill openings that will result as engineers retire or switch to other areas of employment. These openings are expected to account for most of the new jobs in the industry; but job satisfaction and longevity are high in this field, so turnover is usually low.

For More Information

Accreditation Board for Engineering and Technology, Inc.
111 Market Place, Suite 1050
Baltimore, MD 21202
Tel: 410-347-7700
Web: http://www.abet.org

Aerospace Education Foundation
1501 Lee Highway
Arlington, VA 22209
Tel: 703-247-5839
Web: http://www.aef.org

American Institute of Aeronautics and Astronautics
1801 Alexander Bell Drive, Suite 500
Reston, VA 20191-4344
Tel: 703-264-7500
Web: http://www.aiaa.org

American Society for Engineering Education
1818 N Street, NW, Suite 600
Washington, DC 20036
Tel: 202-331-3500
Web: http://www.asee.org

Junior Engineering Technical Society, Inc.
1420 King Street, Suite 405
Alexandria, VA 22314-2794
Tel: 703-548-5387
Web: http://www.jets.org

National Society of Professional Engineers
1420 King Street
Alexandria, VA 22314
Tel: 703-684-2800
Web: http://www.nspe.org

University of North Dakota
International Aerospace Camp
PO Box 9007
Grand Forks, ND 58202-9007
Tel: 701-777-3592
Web: http://www.aero.und.edu

Air Quality Engineers

School Subjects
- Biology
- Chemistry
- Mathematics

Personal Skills
- Communication/ideas
- Technical/scientific

Work Environment
- Primarily indoors
- Primarily one location

Minimum Education Level
- Bachelor's degree

Salary Range
- $23,000 to $35,000 to $70,000+

Certification or Licensing
- Voluntary

Outlook
- Faster than the average

Overview

Air quality engineers, or *air pollution control engineers,* are responsible for developing techniques to analyze and control air pollution by using sophisticated monitoring, chemical analysis, computer modeling, and statistical analysis. Some air quality engineers are involved in pollution-control equipment design or modification. Government-employed air quality experts keep track of a region's polluters, enforce federal regulations, and impose fines or take other action against those who do not comply with regulations. Privately employed engineers may monitor companies' emissions for certain targeted pollutants to ensure that they are within acceptable levels. Air quality engineers who work in research seek ways to combat or avoid air pollution.

History

The growth of cities during the Industrial Revolution was a major contributor to the decline of air quality. Some contaminates (pollutants) have always been with us—for instance, particulate matter (tiny solid particles) from very large fires, or dust caused by wind or mass animal migration. But human populations were not really concentrated enough, nor did the technology exist to produce what is today considered hazardous to the atmosphere, until about 200 years ago. The industrialization of England in the 1750s, followed by France in the 1830s and Germany in the 1850s, changed all that. It created high-density populations as thousands of people were drawn to cities to work in the smoke-belching factories and led to huge increases in airborne pollutants. Work conditions in the factories were notoriously bad, with no pollution-control or safety measures. Rapidly, living conditions in cities became equally bad; the air became severely polluted air and caused respiratory and other diseases.

America's cities were slightly smaller and slower to industrialize, in addition to being more spread out than Old World capitals like London. Even so, levels of sulfur dioxide were so high in Pittsburgh in the early 1900s that ladies' stockings would disintegrate upon prolonged exposure to the air. The rapid growth of the American automobile industry in the first half of the 20th century contributed greatly to air pollution in two ways: initially, from the steel factories and production plants that made economic giants out of places like Pittsburgh and Detroit, and then from the cars themselves. This became an even greater problem as cars enabled people to move out from the fetid industrial city and commute to work there. Mobility independent of public transportation greatly increased auto exhaust and created such modern nightmares as rush hour traffic.

The effects of air pollution were and are numerous. Particulate matter reacts chemically with heat to form ground-level ozone, or smog. Sulfur and nitrogen oxides cause extensive property damage over long periods with their corrosive qualities. Carbon monoxide, the main automobile pollutant, is deadly at a relatively low level of exposure.

According to data from the Environmental Careers Organization, 88 percent of carbon dioxide measured in the Los Angeles basin and 50 percent of that region's volatile organics (a cause of ozone smog) were the result of automobile emissions. This was true despite the Los Angeles basin's having the toughest vehicle emission standards in the United States and the largest market for alternative energy source automobiles. Air pollution affects the environment in well-publicized phenomena like acid rain and holes in the ozone layer, and in less obvious ways as well. For example, a scientist in Great Britain at the turn of the 20th century followed the evolution of white tree moths as natural selection turned them gray to match the birch trees they used for camouflage, which had become covered with a layer of airborne industrial pollutants. Because pollution is so difficult to remove from the air, and because its effects (loss of atmospheric ozone, for example) are so difficult to alter, the problem tends to be cumulative and an increasingly critical public health issue.

Some private air pollution control was implemented in the 20th century, mainly to prevent factories from ruining their own works with corrosive (strongly acid or caustic) and unhealthy emissions. The first attempt at governmental regulation was the Clean Air Act (1955), but because environmental concerns were not considered viable economic or political issues, this act was not very effective. As environmentalists became increasingly visible and vigorous campaigners, the Air Quality Act was established in 1967. The Environmental Protection Agency (EPA) created National Ambient Air Quality Standards (NAAQS) in 1971, which set limits on ozone, carbon monoxide, sulfur dioxide, lead, nitrogen dioxide, and particulate levels in the emissions of certain industries and processes. States were supposed to design and implement plans to meet the NAAQS, but so few complied that Congress was forced to extend deadlines three times. Even now, many goals set by the first generation of air-quality regulations remain unmet, and new pollution issues demand attention. Airborne toxins, indoor air pollution, acid rain, carbon dioxide buildup (the greenhouse effect), and depletion of the ozone are now subjects of international controversy and concern.

The Job

The EPA has composed a list of more than 50 regions of the United States that are out of compliance with federal air quality regulations—some dramatically so—and provided deadlines within the next 20 years to bring these areas under control. The EPA regulations cover everything from car emissions to the greenhouse effect and have the weight of law behind them. There are few industries that will not be touched somehow by this legislation and few that will not require the services of an air quality engineer in the years to come.

Air quality engineers will be the professionals monitoring targeted industries or sources to determine whether they are operating within acceptable emissions levels. These engineers will suggest changes in the setup of specific companies, or even whole industries, to lessen their impact on the atmosphere. There will be ample opportunity in this field to combine interests, precisely because it is a new field with yet unestablished job paths. An air quality engineer with some background in meteorology, for example, might track the spread of airborne pollutants through various weather systems, using computer modeling techniques. Another air quality engineer might research indoor air pollution, discovering causes for the "sick building syndrome" and creating new architectural standards and building codes for safe ventilation and construction materials.

Air quality engineers work for the government, in private industry, as consultants, and in research and development. Government employees are responsible for monitoring a region, citing infractions, and otherwise enforcing government regulations. These workers may be called to give testimony in cases against noncompliant companies. They must deal with public concerns and opinion and are themselves regulated by government bureaucracy and regulations.

Air quality engineers in private industry work within industry or a large company to ensure that air quality regulations are being met. They might be responsible for developing instrumentation to continuously monitor emissions, for example, and using this data to formulate methods of control. They may interact with federal regulators or work independently. Engineers working in private industry also might be involved in what is known as "impact assessment with the goal of sustainable development." This means figuring out the most environmentally sound way to produce products—from raw material to disposal stages—while maintaining or, if possible, increasing the company's profits.

Engineers who work alone as consultants or for consulting firms do many of the same things as engineers in private industry, perhaps for smaller companies who do not need a full-time engineer but still need help meeting federal requirements. They, too, might suggest changes to be implemented by a company to reduce air pollution. Some consultants specialize in certain areas of pollution control. Selling, installing, and running a particular control system is the business of many private consultants. The job requires some salesmanship and the motivation to maintain a variable clientele.

Finally, engineers committed to research and development may work in public or private research institutions and in academic environments. They may tackle significant problems that affect any number of industries and may improve air quality standards with the discovery of new contaminates that need regulation.

Requirements

High School

High school students should develop their skills in chemistry, math, biology, and ecology.

Postsecondary Training

To break into this field, a bachelor's degree in civil, environmental, or chemical engineering is required. Advancement, specialization, or jobs in research may require a master's degree or Ph.D. Besides the regular environmental or chemical engineering curricula at the college level, future air quality engineers might engage in some mechanical or civil engineering if they are interested in product development. Modelers and planners should have a good knowledge of computer systems. Supporting course work in biology, toxicology, or meteorology can give the job seeker an edge for certain specialized positions even before gaining experience in the workforce.

Certification or Licensing

All engineers who do work that affects public health, safety, or property must register with the state. To obtain registration, engineers must have a degree from an accredited engineering program. Right before they get their degree (or soon after), they must pass an engineer-in-training (EIT) exam covering fundamentals of science and engineering. A few years into their careers, engineers also must pass an exam covering engineering practice.

Other Requirements

Prospective air quality engineers should be puzzle solvers. The ability to work with intangibles is a trait of successful air quality management. As in most fields nowadays, communications skills are vital. Engineers must be able to clearly communicate their ideas and findings, both orally and in writing, to a variety of people with different levels of technical understanding.

Exploring

Investigating air quality engineering can begin with reading environmental science and engineering periodicals, available in many large libraries. Familiarizing yourself with the current issues involving air pollution will give you a better idea of what problems will be facing this field in the near future.

The next step might be a call to a local branch of the EPA. In addition to providing information about local source problems, they can also provide a breakdown of air quality standards that must be met and who has to meet them.

To get a better idea about college-level course work and possible career directions, contact major universities, environmental associations, or even private environmental firms. Some private consulting firms will explain how specific areas of study combine to create their particular area of expertise.

Employers

Most air quality engineers are privately employed in industries subject to emissions control, such as manufacturing. They may also work for the federal government, investigating and ensuring compliance with air quality regulations, as consultants to industry and large companies, and in research and development.

Starting Out

Summer positions as an air pollution technician provide valuable insight into the engineer's job as well as contacts and experience. Check with local and state EPA offices and larger consulting firms in your area for internship positions and their requirements. Environmentally oriented engineers may be able to volunteer for citizen watchdog group monitoring programs, patrolling regions for previously undiscovered or unregulated contaminates. Most air quality engineers can expect to get jobs in their field immediately after graduating with a bachelor's degree. Your school placement office can assist you in fine tuning your resume and setting up interviews with potential employers. Government positions are a common point of entry; high turnover rates open positions as experienced engineers leave for the more lucrative private sector, which accounts for four out of five jobs in air quality management. An entry-level job might focus on monitoring and analysis.

Advancement

With experience and education, the engineer might develop a specialization within the field of air quality. Research grants are sometimes available to experienced engineers who wish to concentrate on specific problems or areas of study. Management is another avenue of advancement. The demand for technically oriented middle management in the private sector makes engineers with good interpersonal skills very valuable.

In many ways, advancement will be dictated by the increasing value of air quality engineers to business and industry in general. Successful development of air pollution control equipment or systems—perhaps that even cut costs as they reduce pollution—will make air quality engineers important players in companies' economic strategies. As regulations tighten and increasing emphasis is put on minimizing environmental impact, air quality engineers will be in the spotlight as both regulators and innovators. Advancement may come in the form of monetary incentives or bonuses or management positions over other parts of the organization or company.

Earnings

According to *Engineering News Record,* an estimated $95 billion in revenue will result from growth in this field in the next 20 years. Job opportunities through the next decade will be high, probably higher in those areas of the country targeted by the EPA (generally, larger cities like Los Angeles, Chicago, and Denver). Salaries for entry-level engineers start at around $23,000 to $35,000 per year. Local government agencies pay at the lower end of the scale; state and federal agencies, slightly higher. Salaries in the private sector are highest, from the low $30,000s up to $70,000 or more. Other benefits may include tuition reimbursement programs, use of a company vehicle for fieldwork, full health coverage, retirement plans, and fairly solid job security.

Work Environment

Working conditions differ depending on the employer, the specialization of the position, and the location of the job. An air quality engineer may be required to perform fieldwork, such as observing emission sources, but more often works in an office, determining the factors responsible for airborne pollutants and devising ways to prevent them. Coworkers may include other environmental engineers, lab technicians, and office personnel. An engineer may discuss specific problems with a company's economic planners and develop programs to make that company more competitive environmentally and economically. Those who monitor emissions have considerable responsibility and therefore considerable pressure to do their job well—failure to maintain industry standards could cost their employer government fines. Engineers in some consulting firms may be required to help sell the system they develop or work with.

Most engineers work a standard 40-hour week, putting in overtime to solve critical problems as quickly as possible. A large part of the job for most air quality engineers consists of keeping up-to-date with federal regulations, industry and regional standards, and developments in their area of expertise. Some employers require standard business attire, while some require more fieldwork from their engineers and may not enforce rigorous dress codes. Unlike water and soil pollution, air pollution can sometimes be difficult to measure quantitatively if the source is unknown. Major pollutants are generally easily identified (although not so easily eliminated), but traces of small "leaks" may literally change with the wind and make for time-consuming, deliberate, and frustrating work.

Outlook

When the immediate scramble to modify and monitor equipment slackens as government regulations are met in the next 20 years, the focus in air quality engineering will shift from traditional "end of pipe" controls (e.g., modifying catalytic converters or gasoline to make cars burn gas more cleanly) to source control (developing alternative fuels and eliminating oil-based industrial emissions). As mentioned, impact assessment will play a large part on the corporate side of air quality management, as businesses strive to stay profitable in the wake of public health and safety regulations. Air pollution problems like greenhouse gas buildup and ozone pollution will not be disappearing in the near future and will be increasingly vital areas of research.

International development will allow American pollution control engineers to offer their services in any part of the world that has growing industries or population. Pollution control in general has a big future: air pollution control is quickly becoming a major chunk of the expected expenditures and revenues in this category.

For More Information

The following organization's members work in air pollution control and hazardous waste management:

Air and Waste Management Association
One Gateway Center, Third Floor
Pittsburgh, PA 15222
Tel: 412-232-3444
Email: info@awma.org
Web: http://www.awma.org

The following is an association of pollution control equipment manufacturers:

Institute of Clean Air Companies
660 L Street, NW, Suite 1100
Washington, DC 20036
Tel: 202-457-0911
Web: http://www.icac.com

The following are government pollution control boards:

State and Territorial Air Pollution Program Administrators (STAPPA)
and the Association of Local Air Pollution Control Officials (ALAPCO)
444 North Capitol Street, NW, Suite 307
Washington, DC 20001
Tel: 202-624-7864
Email: 4clnair@sso.org
Web: http://www.4cleanair.org/

Following is the national organization. For your state's Environmental Protection Agency, check the government listings in your phone book:

Environmental Protection Agency
401 M Street, SW
Washington, DC 20460-0003
Tel: 202-260-2090
Web: http://www.epa.gov

Biomedical Engineers

Biology **Chemistry**	School Subjects
Helping/teaching **Technical/scientific**	Personal Skills
Primarily one location **Primarily indoors**	Work Environment
Bachelor's degree	Minimum Education Level
$25,000 to $38,000 to $65,000	Salary Range
Voluntary	Certification or Licensing
Faster than the average	Outlook

Overview

Biomedical engineers are highly trained scientists who use engineering and life science principles to research biological aspects of animal and human life. They develop new theories, and they modify, test, and prove existing theories on life systems. They design health care instruments and devices or apply engineering principles to the study of human systems.

History

Biomedical engineering is one of many new professions created by advancements in technology. It is an interdisciplinary field that brings together two respected professions—biology and engineering.

Biology, of course, is the study of life, and engineering, in broad terms, studies sources of energy in nature and the properties of matter in a way that is useful to humans, particularly in machines, products, and structures. A combination of the two fields, biomedical engineering developed primarily after 1945, as new technology allowed for the application of engineering principles to biology. The artificial heart is just one in a long list of the products of biomedical engineering. Other products include artificial organs, prosthetics, the use of lasers in surgery, cryosurgery and ultrasonics, and the use of computers and thermography in diagnosis.

The Job

Biomedical engineers are employed in industry, hospitals, research facilities of educational and medical institutions, teaching, and government regulatory agencies.

Using engineering principles to solve medical and health-related problems, the biomedical engineer works closely with life scientists, members of the medical profession, and chemists. Most of the work revolves around the laboratory. There are three interrelated work areas: research, design, and teaching.

Biomedical research is multifaceted and broad in scope. It calls upon engineers to apply their knowledge of mechanical, chemical, and electrical engineering as well as anatomy and physiology in the study of living systems. Using computers, biomedical engineers use their knowledge of graphic and related technologies to develop mathematical models that simulate physiological systems.

In biomedical engineering design, medical instruments and devices are developed. Engineers work on artificial organs, ultrasonic imagery devices, cardiac pacemakers, and surgical lasers, for example. They design and build systems that will update hospital, laboratory, and clinical procedures. They also train health care personnel in the proper use of this new equipment.

The teaching aspect of biomedical engineering is on the university level. Teachers conduct classes, advise students, serve on academic committees, and supervise or conduct research.

Within biomedical engineering, an individual may concentrate on a particular specialty area. Some of the well-established specialties are bioinstrumentation, biomechanics, biomaterials, systems physiology, clinical engineering, and rehabilitation engineering. These specialty areas frequently depend on one another.

Biomechanics is mechanics applied to biological or medical problems. Examples include the artificial heart, the artificial kidney, and the artificial hip.

Systems physiology uses engineering strategies, techniques, and tools to gain a comprehensive and integrated understanding of living organisms ranging from bacteria to humans. Biomedical engineers in this specialty examine such things as the biochemistry of metabolism and the control of limb movements.

Rehabilitation engineering is a new and growing specialty area of biomedical engineering. Its goal is to expand the capabilities and improve the quality of life for individuals with physical impairments. Rehabilitation engineers often work directly with the disabled person and modify equipment for individual use.

Requirements

High School

High school students can best prepare for a career as a biomedical engineer by taking courses in biology, chemistry, physics, mathematics, drafting, and computers. Communication and problem-solving skills are necessary, so classes in English, writing, and logic are important. Participation in science clubs and competing in science fairs gives students the opportunity to design and invent systems and products.

Postsecondary Training

Most biomedical engineers have an undergraduate degree in biomedical engineering or a related field and a Ph.D. in some facet of biomedical engineering. Undergraduate study is roughly divided into halves. The first two years are devoted to theoretical subjects, such as abstract physics and differential equations in addition to the core curriculum most undergraduates take. The third and fourth years include more applied science.

During graduate programs, students work on research or product development projects headed by faculty.

Certification or Licensing

Certification and licensing for biomedical engineers is voluntary, and many are not certified. Engineering certification is through the State Board of Technical Registration and requires a five-year residency and passing a comprehensive exam.

Other Requirements

Biomedical engineers should have a strong commitment to learning. They should be scientifically inclined and be able to apply their knowledge in problem solving. Becoming a biomedical engineer requires long years of schooling because a biomedical engineer needs to be an expert in the fields of engineering and biology. Also, biomedical engineers have to be familiar with chemical, material, and electrical engineering as well as physiology and computers.

Exploring

Undergraduate courses offer a great deal of exposure to the field. Working in a hospital where biomedical engineers are employed can also provide insight into the field, as can interviews with practicing or retired biomedical engineers.

Employers

Biomedical engineers are frequently employed in hospitals and medical institutions, and in research and educational facilities. Employment opportunities also exist in industry and government regulatory agencies.

Starting Out

A variety of routes may be taken to gain employment as a biomedical engineer. Recent graduates may use college placement services, or they may apply directly to employers, often to personnel offices in hospitals and industry. A job may be secured by answering an advertisement in the employment section of a newspaper. Information on job openings is also available at the local office of the U.S. Employment Service.

Advancement

Advancement opportunities are tied directly to educational and research background. In a nonteaching capacity, a biomedical engineer with an advanced degree can rise to a supervisory position. In teaching, a doctorate is usually necessary to become a full professor. By demonstrating excellence in research, teaching, and departmental committee involvement, one can move from instructor to assistant professor and then to full professor, department chair, or even dean.

Qualifying for and receiving research grant funding can also be a means of advancing one's career in both the nonteaching and teaching sectors.

Earnings

How much a biomedical engineer earns is dependent upon education and experience. Salaries for biomedical engineers range from $25,000 to $65,000. Those with advanced degrees hold the higher-paying positions, while those with minimum education (bachelor's degree) hold the lower-paying positions.

According to the National Association of Colleges and Employers, beginning salary offers in private industry in 1997 averaged $25,400 a year for those with a bachelor's degree in biological science; about $26,900 for those with a master's degree, and about $52,400 for those with a doctoral degree.

Starting salaries for engineers with the bachelor's degree are significantly higher than starting salaries of bachelor's degree graduates in other fields. According to the National Association of Colleges and Employers, engineering graduates with a bachelor's degree averaged about $38,500 a year in private industry in 1997; those with a master's degree and no experience, $45,400 a year; and those with a Ph.D., $59,200.

In colleges and universities in the 1990s, salaries for full-time faculty members on nine-month contracts ranged from about $26,800 for instructors to $63,000 for professors. University professors can supplement their income significantly by writing and consulting.

Earnings in the private sector generally run higher than those in government or education. The average annual salary for engineers in the federal government in nonsupervisory, supervisory, and managerial positions was $61,950 in 1997.

Work Environment

Those engaged in university teaching will have much student contact in the classroom, the laboratory, and the office. They also will be expected to serve on relevant committees while continuing their teaching, research, and writing responsibilities. As competition for teaching positions increases, the requirement that professors publish papers will increase. Professors usually are responsible for obtaining government or private research grants to support their work.

Outlook

In the 1990s there were more than 4,000 biomedical engineers in the United States. They were employed in all parts of the country in hospitals, colleges and universities, medical and engineering schools, federal and state agencies, and private industry.

It is expected that there will be a greater need for skilled biomedical engineers in the future. Prospects look particularly good in the large health care industry, which will continue to grow rapidly, primarily because people are living longer. New jobs will become available in biomedical research in prosthetics, artificial internal organs, computer applications, and instrumentation and other medical systems. In addition, a demand will exist for teachers to train the biomedical engineers needed to fill these positions.

For More Information

For more information on careers in biomedical engineering, contact the following:

American Society for Engineering Education
1818 North N Street, NW, Suite 600
Washington, DC 20036
Tel: 202-331-3500
Web: http://www.asee.org

Biomedical Engineering Society
PO Box 2399
Culver City, CA 90231
Tel: 310-618-9322
Email: ames@netcom.com
Web: http://www.bme.ecn.purdue.edu/bme

Canadian Medical and Biological Engineering Society
National Research Council of Canada
Room 393, M-1500
Ottawa ON K1A 0R8 Ontario
Canada

Ceramics Engineers

Chemistry Physics	School Subjects
Technical/scientific Mechanical/manipulative	Personal Skills
Bachelor's degree	Minimum Education Level
$35,000 to $59,000 to $99,000	Salary Range
None available	Certification or Licensing
Little change or more slowly than the average	Outlook

Overview

Ceramics engineers work with nonmetallic elements such as clay and inorganic elements such as zirconia. Numbering fewer than 20,000 in the United States, they are part of the ceramics and glass industry, which manufactures such common items as tableware and such highly technical items as ceramic tiles for the space shuttle. They perform research, design machinery and processing methods, and develop new ceramic materials and products. They work at engineering consulting firms, manufacturing and industrial plants, and laboratories.

History

When we refer to ceramics we often think only of objects made of clay, like cups and saucers. Thousands of years ago, ceramics makers were limited by a dependence on this one raw material. Originally, clay was probably merely dried in the sun to harden before use. By seven thousand years ago it was being fired to make it more durable, but not many further advancements were made in its development and use for thousands of years. Ceramics makers produced things made of clay to fulfill the basic household need of storing and serving food and liquids. Based on research, it is believed that making pottery was exclusively the work of women. Throughout the world and over the centuries, changes made by workers were relatively minor when considering the many uses of ceramics today; they changed basic forms and glazings, their artistic impressions developed and spread, and they used higher-temperature materials.

Not until the scientific and industrial revolutions of the 19th century did ceramics begin to have a place in higher technology. Engineers became concerned with materials that were inorganic and nonmetallic and had industrial applications. The earliest industrial ceramics engineers used porcelains for high-voltage electrical insulation. Ceramics engineers benefited other industries as well, developing, for example, material for spark plugs (automotive and aerospace industries), magnetic materials (electronics industry), and nuclear fuel rods (nuclear industry).

Basic ceramics materials such as clay and sand are still being used by artists and craftspeople. But in the important, although perhaps obscure, field of industrial ceramics, they are used mainly in mechanical engineering and electrical/electronics engineering applications. Ceramics engineers are working with more advanced materials as well (many produced by chemical processes), including high-strength silicon carbides, nitrides, and fracture-resistant zirconias.

The Job

Like other materials engineers, ceramics engineers work toward the development of new products. They also use their scientific knowledge to anticipate new applications for existing products.

Ceramics research engineers conduct experiments and perform other research. They study the chemical properties (such as sodium content) and physical properties (such as heat resistance) of materials as they develop the ideal mix of elements for each product's application. Many research engineers are fascinated by the chemical and thermal interactions of the oxides that make up many ceramic materials.

Ceramics design engineers take the information culled by the researchers and further develop actual products to be tested. They also design equipment used in ceramics manufacturing, such as grinders, milling machines, sieves, presses, and drying machines.

Ceramics test engineers test materials that have been chosen by the researchers to be used as sample products, or they might be involved in ordering raw materials and making sure the quality meets the ceramics industry standards. Other ceramics engineers are involved in more hands-on work, such as grinding raw materials and firing products. Maintaining proper color, surface finish, texture, strength, and uniformity are further tasks that are the responsibility of the ceramics engineer.

Beyond research, design, testing, and manufacturing, there are the *ceramics products sales engineers.* The industry depends on these people to anticipate customers' needs and report back to researchers and test engineers on new applications.

Ceramics engineers often specialize in one or a few areas, including the following (which are associated with selected products in each niche): glass (windows, light bulbs, tableware, fiber optics, electronic equipment parts); structural clay (sewer pipe, enameled cookware, jewelry); cement (for building materials); and whitewares (pottery, china, wall tile, plumbing fixtures, electrical insulators, spark plugs).

A relatively new segment of the industry—advanced, or technical, ceramics—employs even more specialized workers. In this part of the field, workers are involved in either structural ceramics (for things like engine components, cutting tools, military armor, and replacement parts for the body) or electrical ceramics (for products like integrated circuits, sensors, and magnets).

Requirements

High School

If you are interested in this field, no doubt you like to ask questions about how materials work and how elements react to each other. What makes concrete crackable? What makes fiber optics carry messages over space? Ceramics engineers have inquiring minds, often analyzing and trying to figure things out.

Ceramics engineering involves learning a lot of scientific material. In high school, science classes are the key—physics and chemistry in particular. However, all other basic courses need to be concentrated on as well: English, math, history, and social science.

High school will not be your last stop for education. You will need at least a bachelor's degree to get a job in ceramics engineering.

Postsecondary Training

Your first college courses will initially be geared toward getting you to think logically and analytically. Thus, the first two years of engineering programs typically center on math, physics, chemistry, and computer courses. You should be inspired and challenged to approach problems first theoretically and then practically. For instance, after you are presented with a problem, you will first think about how it would be solved, then formulate a step-by-step method by which to solve it, and then actually tackle the problem according to that method. This thinking process should be nurtured throughout your core college courses geared toward ceramics engineering, for engineers are expected to have such an aptitude for problem solving.

In your junior and senior years, you will focus particularly on your chosen area of specialization. If you major in ceramics engineering, classes and problems will be concentrated on issues in the industrial ceramics engineering discipline. During these last two years of undergraduate work, it is important to consider and evaluate your goals in the field and to determine whether you prefer research, production, design, sales, or management. Focusing on this objective makes it less difficult for you to plan your job search.

Other Requirements

As mentioned earlier, you should be an inquisitive person with somewhat of an analytical mind. Since you might be doing testing and recording of process results, you'll need to be relatively comfortable working with details. You should enjoy doing intellectually demanding work and be disciplined and motivated enough to do your job without close and constant supervision, paying close attention to what you are doing. It's also important that you be able to communicate well and get along with your co-workers because engineers often work together.

Exploring

If you're interested in ceramics engineering, it's a good idea to take on special research assignments from teachers who can provide guidance on topics and methods. There are also summer academic programs where students with similar interests can spend a week or more in a special environment. The Worcester Polytechnic Institute in Worcester, Massachusetts, for example, offers a 13-day program focusing on science, math, and engineering (sports such as swimming and softball are also offered). It's also a good idea to join a national science club, such as the Junior Engineering Technical Society. In this organization, member students have the opportunity to compete in academic events, take career exploration tests, and enter design contests where they build models of such things as spacecraft and other structures based on their own designs.

For hands-on experience with materials, take pottery or sculpture classes, and this will allow you to become familiar with things like clay and glass. You can learn how the materials are obtained and how to shape and fire them. This will give you an opportunity to learn firsthand about stress and strain, tension and compression, heat resistance, and ideal production equipment. In pottery classes, you can also learn about glazes and how various chemicals affect different materials.

Employers

Many different kinds of employers hire ceramics engineers for the variety of positions they can fill. They are involved at manufacturing companies, electronics industries, and research and testing organizations. They work in industries that produce and process metal; machinery; electrical equipment; aircraft; and stone, clay, and glass products. They work in chemical and mining industries that make and use ceramic products, as well as in computer and semiconductor industries. They also work in federal government agencies and engineering consulting companies.

Starting Out

As a high school senior, you might want to inquire with established manufacturing companies about internships and summer employment opportunities. College placement centers can also help you find employers that participate in cooperative education programs, where high school students work at a materials engineering job in exchange for course credits.

Companies looking for engineers often send recruiters to colleges to talk with students, so when you are a senior in college you should register with your school's placement office. You must have at least a bachelor's degree to get an entry-level job in ceramics engineering. In your first job you will often be working as part of a team supervised by experienced engineers, or you might first go through a company training program. As a new employee, you also might find yourself assisting researchers in the lab.

Advancement

As you continue to gain research experience, you can apply for higher-level jobs in production and marketing. After becoming familiar with the materials and products at your company or organization, you may be assigned to on-the-job training in production areas. You might even go on to work as a supervisor. Some engineers decide to transfer to teaching positions at universities after working a number of years in the industry.

Opportunities for advancement are available especially for those who continue their education throughout their work years. Technology is always advancing, and new products and applications continue to be developed, so if you keep up-to-date on issues in materials science you are more likely to succeed. Some engineers leave the field after many years to take top-level management positions in other industries. Those who are employed by universities must continue teaching if they want to gain tenured (that is, further guaranteed) positions.

Earnings

According to the 1998-99 *Occupational Outlook Handbook* information on engineering in general, earnings for ceramics engineers are considered quite good. The average annual salary for a ceramics engineer just beginning a career would be about $35,000. Those who have been working in the field for some years earn an average of about $59,000; those in senior managerial positions with years of experience can earn nearly $100,000 per year. Salaries for government workers are generally less than those who work for private companies.

The fringe benefits that most ceramics engineers earn are similar to those in many other industries: health insurance, sick leave, paid holidays, pension plans, and paid vacations.

Work Environment

Working conditions in ceramics engineering positions vary depending on the specific field and department in which one works. Hands-on engineers work in plants and factories. Researchers work mainly in laboratories, research institutes, and universities. Those in management positions work mostly in offices; and teachers, of course, work in school environments.

Whatever your job description, you'll probably be working a standard eight-hour day, five days a week. You'll be indoors, either in an office, a research lab, a classroom, or a factory. If you were a ceramics research engineer, for example, you might be in a lab conducting studies on the properties of ceramics and how processing affects them. You could be testing things like materials' texture, color, heat resistance, electrical resistivity, or corrosiveness. If you were a production engineer, you might be involved in developing the equipment needed by a factory for making products that include some type of ceramic. Maybe you'll be discovering new abrasives for grinding metal or designing medical instruments or artificial body parts. In any case, your scientific knowledge and analytical capabilities will continue to be challenged.

Outlook

Although employment for ceramics engineers is expected to grow more slowly than the average through 2006 (according to the 1998-99 *Occupational Outlook Handbook*), you shouldn't find a lot of competition for jobs because of the low number of new graduates these days in ceramics engineering. And even though industries that need ceramics engineers (those specifically involved in stoneware, clay, and glass products; primary metals; fabricated metal products; and transportation equipment) are not expected to have much growth in the near future, service segments of the ceramics industry are expected to grow; these sections include research and testing and engineering and architectural services. Ceramics engineers are expected to be needed to develop improved materials for the industrial customers of these firms.

For More Information

The following international association provides the latest technical, scientific, and educational information to its members and others in the ceramics and related materials field.

American Ceramic Society
735 Ceramic Place, PO Box 6136
Westerville, OH 43086-6136
Tel: 614-794-5898
Web: http://www.acers.org

JETS provides activities, events, competitions, programs, and materials that allow high school students to "try on" engineering. Visit the JETS homepage for a complete list of programs, including a broad range of guidance materials about the various engineering disciplines.

Junior Engineering Technical Society, Inc.
1420 King Street, Suite 405
Alexandria, VA 22314-2794
Tel: 703-548-5387
Email: jets@nae.edu
Web: http://www.jets.org

TMS is a society committed to promoting the global science and engineering professions concerned with minerals, metals, and materials.

The Minerals, Metals & Materials Society (TMS)
420 Commonwealth Drive
Warrendale, PA 15086-7514
Tel: 724-776-9000
Web: http://www.tms.org

This is a program for high school seniors that covers science material not traditionally offered in high school.

Worcester Polytechnic Institute
Frontiers in Science, Mathematics, and Engineering
100 Institute Road
Worcester, MA 01609-2280
Tel: 508-831-5514
Web: http://www.wpi.edu

Chemical Engineers

Chemistry Physics	School Subjects
Communication/ideas Technical/scientific	Personal Skills
Primarily indoors Primarily one location	Work Environment
Bachelor's degree	Minimum Education Level
$40,000 to $75,000 to $125,000	Salary Range
Recommended	Certification or Licensing
About as fast as the average	Outlook

Overview

Chemical engineers take chemistry out of the laboratory and into the real world. They are involved in evaluating methods and equipment for the mass production of chemicals and other materials requiring chemical processing. They also develop products from these materials, such as plastics, metals, gasoline, detergents, pharmaceuticals, and foodstuffs. They develop or improve safe, environmentally sound processes; determine the least costly production method; and formulate the material for easy use and safe, economic transportation.

History

Chemical engineering, defined in its most general sense as applied chemistry, existed even in early civilizations. Ancient Greeks, for example, distilled alcoholic beverages, as did the Chinese, who by 800 BC had learned to distill alcohol from the fermentation of rice. Aristotle, a fourth-century Greek philosopher, wrote about a process for obtaining fresh water by evaporating and condensing water from the sea.

The foundations of modern chemical engineering were laid out during the Renaissance, when experimentation and the questioning of accepted scientific theories became widespread. This period saw the development of many new chemical processes, such as those for producing sulfuric acid (for fertilizers and textile treatment) and alkalis (for soap). The atomic theories of John Dalton and Amedeo Avogadro, developed in the 1800s, supplied the theoretical underpinning for modern chemistry and chemical engineering.

With the advent of large-scale manufacturing in the mid-19th century, modern chemical engineering began to take shape. Chemical manufacturers were soon required to seek out chemists familiar with manufacturing processes. These early chemical engineers were called chemical technicians or industrial chemists. The first course in chemical engineering was taught in 1888 at the Massachusetts Institute of Technology, and by 1900, "chemical engineer" had become a widely used job title.

Chemical engineers are employed in increasing numbers to design new and more efficient ways to produce chemicals and chemical by-products. In the United States, they have been especially important in the development of petroleum-based fuels for automotive vehicles. Their achievements range from the large-scale production of plastics, antibiotics, and synthetic rubbers to the development of high-octane gasoline.

The Job

Chemical engineering is one of the four major engineering disciplines (the others are electrical, mechanical, and civil). Because chemical engineers are rigorously trained not only in chemistry but also in physics, mathematics, and other sciences such as biology or geology, they are among the most versatile of all engineers, with many specialties, and are employed in many industries. Chemical industries, which transform raw materials into desired products, employ the largest number of chemical engineers.

Research engineers work with chemists to develop new processes and products, or they may develop better methods to make existing products. Product ideas may originate with the company's marketing department; with a chemist, chemical engineer, or other specialist; or with a customer. The basic chemical process for the product is then developed in a laboratory, where various experiments are conducted to determine the process's viability. Some projects die here.

Others go on to be developed and refined at pilot plants, which are small-scale versions of commercial plants. Chemical engineers in these plants run tests on the processes and make any necessary modifications. They strive to improve the process, reduce safety hazards and waste, and cut production time and costs. Throughout the development stage, engineers keep detailed records of the proceedings—and may abandon projects that aren't viable.

When a new process is judged to be viable, *process design engineers* determine how the product can most efficiently be produced on a large scale while still guaranteeing a consistently high-quality result. These engineers consider process requirements and cost, convenience and safety for the operators, waste minimization, legal regulations, and preservation of the environment. Besides working on the steps of the process, they also work on the design of the equipment to be used in the process. These chemical engineers are often assisted in plant and equipment design by mechanical, electrical, and civil engineers.

Project engineers oversee the construction of new plants and installation of new equipment. In construction, chemical engineers may work as *field engineers,* who are involved in the testing and initial operation of the equipment and assist in plant start-up and operator training. Once a process is fully implemented at a manufacturing plant, *production engineers* supervise the day-to-day operations. They are responsible for the rate of production, scheduling, worker safety, quality control, and other important operational concerns.

Chemical engineers working in environmental control are involved in waste management, recycling, and control of air and water pollution. They work with the engineers in research and development, process design, equipment and plant construction, and production to incorporate environmental protection measures into all stages of the chemical engineering process.

As *technical sales engineers,* chemical engineers may work with customers of manufactured products to determine what best fits their needs. They answer questions such as, "Could our products be used more economically than those now in use? Why does this paint peel?" etc. Others work as managers, making policy and business decisions and overseeing the training of new personnel. Still others may act as *biomedical engineers* who work with physicians to develop systems to track critical chemical processes in the body or look for the best method of administering a particular drug to a patient. The variety of job descriptions is almost limitless because of chemical engineers' versatility and adaptability.

Requirements

High School

High school students interested in chemical engineering should take all the mathematics and science courses their schools offer. These should include algebra, geometry, calculus, trigonometry, chemistry, physics, and biology. Computer science courses are also highly recommended. In addition, students should take four years of English, and a foreign language is valuable. To enhance their desirability, students should participate in high school science and engineering clubs and other extracurricular activities.

Postsecondary Training

A bachelor's degree in chemical engineering is the minimum educational requirement for entering the field. For some positions, an M.S., an M.B.A., or a Ph.D. may be required. A Ph.D. may be essential for advancement in research, teaching, and administration.

For their college studies, students need a chemical engineering program approved by the Accreditation Board for Engineering and Technology and the American Institute of Chemical Engineers. There are about 145 accredited undergraduate programs in chemical engineering in the United States offering bachelor's degrees. Some engineering programs last five or six years; these often include work experience in industry.

As career plans develop, the student should consult with advisors about special career paths in which they are interested. Those who want to teach or conduct research will need a graduate degree. There are approximately 140 accredited chemical engineering graduate programs in the United States. A master's degree generally takes two years of study beyond undergraduate school, while a Ph.D. program requires four to six years.

In graduate school, students specialize in one aspect of chemical engineering, such as chemical kinetics or biotechnology. Graduate education also helps to obtain promotions, and some companies offer tuition reimbursement to encourage employees to take graduate courses. For engineers who would like to become managers, a master's degree in business administration may be helpful. Chemical engineers must be prepared for a lifetime of education to keep up with the rapid advances in technology.

Certification or Licensing

Chemical engineers must be licensed if they wish to work in the public sector. All 50 states and the District of Columbia have specific licensing requirements, which include graduation from an accredited engineering school, passing a written exam, and having at least four years of engineering experience. About one-third of all chemical engineers are licensed; they are called *registered engineers.*

Other Requirements

Important personal qualities are honesty, accuracy, objectivity, and perseverance. In addition, chemical engineers must be inquisitive, open-minded, creative, and flexible. Problem-solving ability is essential. To remain competitive in the job market, they should display initiative and leadership skills and exhibit the ability to work well in teams, collaborate across disciplines, and be able to work with people of different languages and cultures.

Exploring

High school students should join science clubs and take part in other extracurricular activities and such organizations as the Junior Engineering Technical Society (JETS). JETS participants have opportunities to enter engineering design and problem-solving contests and to learn team development skills. Science contests are also a good way to apply principles learned in classes to a special project. Students can also subscribe to the American Chemical Society's *Chem Matters,* a quarterly magazine for high school chemistry students.

College students can join professional associations, such as the American Chemical Society, the American Institute of Chemical Engineers, and the Society of Manufacturing Engineers (composed of individual associations with specific fields of interest), as student affiliates. Membership benefits include subscription to magazines—some of them geared specifically toward students—that provide the latest industry information. College students can also contact ACS or AIChE local sections to arrange to talk with some chemical engineers about what they do. These associations can also help them find summer or co-op work experiences.

In addition, the Society of Women Engineers has a mentor program in which high school and college women are matched with a SWE member in their area. This member is available to answer questions and provide a first-hand introduction to a career in engineering.

Employers

While the bulk of chemical engineers (about 42 percent) work in chemical industries, the remainder are employed by a variety of industries, the federal and state governments, and colleges and universities. The list of individual employers, if cited, would take many pages. However, the following industry classifications indicate where most chemical engineers are employed: fuels, electronics, food and consumer products, design and construction, materials, aerospace, biotechnology, pharmaceuticals, environmental control, pulp and paper, public utilities, and consultation firms. Because of the nature of their training and background, they can easily obtain employment with another company in a completely different field if necessary or desired.

Starting Out

Most chemical engineers obtain their first position through company recruiters sent to college campuses. Others may find employment with companies with whom they have had summer or work study arrangements. Many respond to advertisements in professional journals or newspapers. The Internet now offers multiple opportunities to job seekers, and many libraries have programs that offer assistance in making use of the available job listings. Chemical engineers may also contact colleges and universities regarding positions as part-time teaching or laboratory assistants if they wish to continue study for a graduate degree. Student members of professional societies often use the employment services of these organizations, including resume data banks, online job listings, national employment clearing houses, and employers' mailing lists.

Typically, new recruits begin as trainees or process engineers. They often begin work under the supervision of seasoned engineers. Many participate in special training programs designed to orient them to company processes, procedures, policies, and products. This allows the company to determine where the new personnel may best fulfill their needs. After this training period, new employees often rotate positions to get an all-around experience in working for the company.

Advancement

Entry-level personnel usually advance to project or production engineers after learning the ropes in product manufacturing. They may then be assigned to sales and marketing. A large percentage of engineers no longer do engineering work by the tenth year of their employment. At that point, they often advance to supervisory or management positions. An MBA enhances their opportunities for promotion. A doctoral degree is essential for university teaching or supervisory research positions. Some engineers may decide at this point that they prefer to start their own consulting firms. Continued advancement, raises, and increased responsibility are not automatic, but depend upon sustained demonstration of leadership skills.

Earnings

Chemical engineering is one of the highest paid scientific professions. Salaries vary with education, experience, industry, and employer. A survey of data for 1997 by the American Institute of Chemical Engineers (AIChE) found median starting salaries as follows: B.S. $42,000; M.S. $46,500; and Ph.D. $63,000. Median salaries for full-time chemical engineers employed in industry were: B.S. $63,000; M.S. $75,000; and Ph.D. $80,000, with the highest median salaries in the area of petroleum engineers. For chemical engineers with doctoral degrees and many years of experience who attain supervisory and management positions, salaries above $100,000 are not unusual. Median salaries for all chemical engineers employed by the federal government were $67,000, and by educational institutions, $78,000.

Work Environment

Because the industries in which chemical engineers work are so varied—from academia to waste treatment and disposal—the working conditions also vary. Most chemical engineers work in clean, well-maintained offices, laboratories, or plants, although some occasionally work outdoors, particularly construction engineers. Travel to new or existing plants may be required. Some chemical engineers work with dangerous chemicals, but the adoption of safe working practices has greatly reduced potential health hazards. Chemical engineers at institutions of higher learning spend their time in classrooms or research laboratories.

The workweek for a chemical engineer in manufacturing is usually 40 hours, although many work longer hours. Because plants often operate around the clock, they may work different shifts or have irregular hours.

Outlook

According to a 1998 report by the Bureau of Labor Statistics, employment in the chemical manufacturing industry is projected to grow slowly through 2006 with an approximately 20 percent growth rate. Employment of chemical engineers should increase about as fast as the average for all occupations as chemical companies research and develop new chemicals and increase the efficiency of their production. Chemical engineering graduates may face competition for engineering jobs in manufacturing as the number of these openings is projected to be lower than the number of graduates. This should not present a problem, however, since due to the broad technical base of chemical engineering education, companies in many related fields are glad to add chemical engineering graduates to their rosters.

For More Information

In addition to general career information and a job bank, ACS offers a directory of experience opportunities that lists undergraduate internships, summer jobs, and co-op programs.

American Chemical Society (ACS)
Career Education
1155 16th Street, NW
Washington, DC 20036
Tel: 202-452-2113
Web: http://www.acs.org

For a chemical engineering career packet, contact:

American Institute of Chemical Engineers (AIChE)
Communications Department
345 East 47th Street
New York, NY 10017-2395
Tel: 212-705-7660 or 800-242-4363
Web: http://www.aiche.org

For information about JETS programs, products, and a chemical engineering career brochure, send a stamped, self-addressed envelope to

Junior Engineering Technical Society, Inc.
1420 King Street, Suite 405
Alexandria, VA 22314-2794
Tel: 703-548-5387
Web: http://www.jets.org

For information on SME associations or on how to become a student member, contact:

Society of Manufacturing Engineers
Professional Interests Department
One SME Drive
PO Box 930
Dearborn, MI 48121
Tel: 313-271-1500
Web: http://www.sme.org

Civil Engineers

	School Subjects
Mathematics Physics	

	Personal Skills
Leadership/management Technical/scientific	

	Work Environment
Indoors and outdoors Primarily multiple locations	

	Minimum Education Level
Bachelor's degree	

	Salary Range
$33,000 to $46,000 to $117,000	

	Certification or Licensing
Required by all states	

	Outlook
Faster than the average	

Overview

Civil engineers are involved in the design and construction of the physical structures that make up our surroundings, such as roads, bridges, buildings, and harbors. Civil engineering involves theoretical knowledge applied to the practical planning of the layout of our cities, towns, and other communities. It is concerned with modifying the natural environment and building new environments to better the lifestyles of the general public. Civil engineers are also known as *structural engineers*.

History

One might trace the evolution of civil engineering methods by considering the building and many reconstructions of England's London Bridge. In Roman and medieval times, several bridges made of timber were built over the Thames River. Around the end of the 12th century, these were rebuilt into nineteen narrow arches mounted on piers. A chapel was built on one of the piers, and two towers were built for defense. A fire damaged the bridge around 1212, yet the surrounding area was considered a preferred place to live and work, largely because it was the only bridge over which one could cross the river. The structure was rebuilt many times during later centuries using different materials and designs. By 1830, it had only five arches. More than a century later, the center span of the bridge was remodeled, and part of it was actually transported to the United States to be set up as a tourist attraction.

Working materials for civil engineers have changed during many centuries. For instance, bridges, once made of timber, then of iron and steel, are today made mainly with concrete that is reinforced with steel. The high strength of the material is necessary because of the abundance of cars and other heavy vehicles that travel over the bridges.

As the population continues to grow and communities become more complex, structures that civil engineers must pay attention to have to be remodeled and repaired. New highways, buildings, airstrips, and so forth must be designed to accommodate public needs. Today, more and more civil engineers are involved with water treatment plants, water purification plants, and toxic waste sites. Increasing concern about the natural environment is also evident in the growing number of engineers working on such projects as preservation of wetlands, maintenance of national forests, and restoration of sites around land mines, oil wells, and industrial factories.

The Job

Civil engineers use their knowledge of materials science, engineering theory, economics, and demographics to devise, construct, and maintain our physical surroundings. They apply their understanding of other branches of science—such as hydraulics, geology, and physics—to design the optimal blueprint for the project.

Feasibility studies are conducted by *surveying and mapping engineers* to determine the best sites and approaches for construction. They extensively investigate the chosen sites to verify that the ground and other surroundings are amenable to the proposed project. These engineers use sophisticated equipment, such as satellites and other electronic instruments, to measure the area and conduct underground probes for bedrock and groundwater. They determine the optimal places where explosives should be blasted in order to cut through rock.

Many civil engineers work strictly as consultants on projects, advising their clients. These consultants usually specialize in one area of the industry, such as water systems, transportation systems, or housing structures. Clients include individuals, corporations, and the government. Consultants will devise an overall design for the proposed project, perhaps a nuclear power plant commissioned by an electric company. They will estimate the cost of constructing the plant, supervise the feasibility studies and site investigations, and advise the client on whom to hire for the actual labor involved. Consultants are also responsible for such details as accuracy of drawings and quantities of materials to order.

Other civil engineers work mainly as contractors and are responsible for the actual building of the structure; they are known as *construction engineers*. They interpret the consultants' designs and follow through with the best methods for getting the work done, usually working directly at the construction site. Contractors are responsible for scheduling the work, buying the materials, maintaining surveys of the progress of the work, and choosing the machines and other equipment used for construction. During construction, these civil engineers must supervise the labor and make sure the work is completed correctly and efficiently. After the project is finished, they must set up a maintenance schedule and periodically check the structure for a certain length of time. Later, the task of ongoing maintenance and repair is often transferred to local engineers.

Civil engineers may be known by their area of specialization. *Transportation engineers,* for example, are concerned mainly with the construction of highways and mass transit systems, such as subways and commuter rail lines. When devising plans for subways, engineers are responsible for considering the tunneling that is involved. *Pipeline engineers* are specialized civil engineers who are involved with the movement of water, oil, and gas through miles of pipeline.

Requirements

High School

Because a bachelor's degree is considered essential in the field, high school students interested in civil engineering must follow a college prep curriculum. Students should focus on mathematics (algebra, trigonometry, geometry, and calculus), the sciences (physics and chemistry), computer science, and English and the humanities (history, economics, and sociology). Students should also aim for honors-level courses.

Postsecondary Training

In addition to completing the core engineering curriculum (including mathematics, science, drafting, and computer applications), students can choose their specialty from the following types of courses: structural analysis; materials design and specification; geology; hydraulics; surveying and design graphics; soil mechanics; and oceanography. Bachelor's degrees can be achieved through a number of programs: a four- or five-year accredited college or university; two years in a community college engineering program plus two or three years in a college or university; five or six years in a co-op program (attending classes for part of the year and working in an engineering-related job for the rest of the year). About 30 percent of civil engineering students go on to receive a master's degree.

Certification or Licensing

Most civil engineers go on to study and qualify for a Professional Engineer (P.E.) license. It is required before one can work on projects affecting property, health, or life. Because many engineering jobs are found in government specialties, most engineers take the necessary steps to obtain the license. Registration guidelines are different for each state—they involve educational, practical, and teaching experience. Applicants must take an examination on a specified date.

Other Requirements

Basic personal characteristics often found in civil engineers are an avid curiosity; a passion for mathematics and science; an aptitude for problem solving, both alone and with a team; and an ability to visualize multidimensional, spatial relationships.

Exploring

High school students can become involved in civil engineering by attending a summer camp or study program in the field. For example, the Worcester Polytechnic Institute in Massachusetts has a twelve-day summer program for students in junior and senior high school. Studies and events focus on science and math and include specialties for those interested in civil engineering.

After high school, another way to learn about civil engineering duties is to work on a construction crew that is involved in the actual building of a project designed and supervised by engineers. Such hands-on experience would provide an opportunity to work near many types of civil workers. Try to work on highway crews or even in housing construction.

Starting Out

To establish a career as a civil engineer, one must first receive a bachelor's degree in engineering or another appropriate scientific field. College placement offices are often the best sources of employment for beginning engineers. Entry-level jobs usually involve routine work, often as a member of a supervised team. After a year or more (depending on job performance and qualifications), one becomes a junior engineer, then an assistant to perhaps one or more supervising engineers. Establishment as a Professional Engineer comes after passing the P.E. exam.

Advancement

Professional Engineers with many years' experience often join with partners to establish their own firms in design, consulting, or contracting. Some leave long-held positions to be assigned as top executives in industries such as manufacturing and business consulting. Also, there are those who return to academia to teach high school or college students. For all of these potential opportunities, it is necessary to keep abreast of engineering advancements and trends by reading industry journals and taking courses.

Earnings

Civil engineers are among the lowest paid in the engineering field. However, starting salaries are usually higher than for other occupations. Entry-level civil engineers with a bachelor's degree earn approximately $33,000 per year in private industry; those with a master's degree, about $35,000; and those with a doctorate, about $47,000. Those working in government jobs earn less than civil engineers at private companies. As with all occupations, salaries are higher for those with more experience. The average salary for those in mid-level positions is $46,000 in private industry and $61,000 in government jobs. Top civil engineers earn as much as $100,000 a year.

Work Environment

Many civil engineers work regular 40-hour weeks, often in or near major industrial and commercial areas. Sometimes they are assigned to work in remote areas and foreign countries. Because of the diversity of civil engineering positions, working conditions vary widely. Offices, labs, factories, and actual sites are typical environments for engineers. About 40 percent of all civil engineers can be found working for various levels of government, usually involving large public-works projects, such as highways and bridges.

A typical work cycle involving various types of civil engineers involves three stages: planning, constructing, and maintaining. Those involved with development of a campus compound, for example, would first need to work in their offices developing plans for a survey. Surveying and mapping engineers would have to visit the proposed site to take measurements and perhaps shoot aerial photographs. The measurements and photos would have to be converted into drawings and blueprints. Geotechnical engineers would dig wells at the site and take core samples from the ground. If toxic waste or unexpected water is found at the site, the contractor determines what should be done.

Actual construction then begins. Very often, a field trailer on the site becomes the engineers' makeshift offices. The campus might take several years to build—it is not uncommon for engineers to be involved in long-term projects. If contractors anticipate that deadlines will not be met, they often put in weeks of 10- to 15-hour days on the job.

After construction is complete, engineers spend less and less time at the site. Some may be assigned to stay on-site to keep daily surveys of how the structure is holding up and to solve problems when they arise. Eventually, the project engineers finish the job and move on to another long-term assignment.

Outlook

Through the year 2006, civil engineers are expected to experience steady employment for the maintenance and repair of public works, such as highways and water systems. The need for civil engineers will depend somewhat on the government's decisions to spend further on renewing and adding to the country's basic infrastructure. As public awareness of environmental issues continues to increase, civil engineers will find expanding employment opportunities at wastewater sites, recycling establishments, and toxic dump sites for industrial and municipal waste.

For More Information

For information on careers and scholarships, contact:

American Society of Civil Engineers
1801 Alexander Bell Drive
Reston, VA 20191-4400
Tel: 703-295-6000
Web: http://www.asce.org

For information on careers, scholarships, and a list of accredited schools, contact:

Institute of Transportation Engineers
525 School Street, SW, Suite 410
Washington, DC 20024
Tel: 202-554-8050
Web: http://www.ite.org

For information on careers in engineering, contact:

Junior Engineering Technical Society, Inc.
1420 King Street, Suite 405
Alexandria, VA 22314
Tel: 703-548-5387
Web: http://www.jets.org

Electrical and Electronics Engineers

School Subjects
Computer science
Physics
Mathematics

Personal Skills
Mechanical/manipulative
Technical/scientific

Work Environment
Primarily indoors
One location with some travel

Minimum Education Level
Bachelor's degree

Salary Range
$39,500 to $50,000 to $100,000+

Certification or Licensing
Voluntary

Outlook
Faster than the average

Overview

Electrical engineers apply their knowledge of the sciences to working with equipment that produces and distributes electricity, such as generators, transmission lines, and transformers. They also design, develop, and manufacture electric motors, electrical machinery, and ignition systems for automobiles, aircraft, and other engines. *Electronics engineers* are more concerned with devices made up of electronic components such as integrated circuits and microprocessors. They design, develop, and manufacture products such as computers, telephones, and radios. Electronics engineering is a subfield of electrical engineering, and both types of engineers are often referred to as electrical engineers. There are approximately 367,000 electrical and electronics engineers employed in the United States.

History

Electrical and electronics engineering had their true beginnings in the 19th century. In 1800, Alexander Volta (1745-1827) made a discovery that opened a door to the science of electricity—he found that electric current could be harnessed and made to flow. By the mid-1800s the basic rules of electricity were established, and the first practical applications appeared. At that time, Michael Faraday (1791-1867) discovered the phenomenon of electromagnetic induction. Further discoveries followed. In 1837 Samuel Morse (1791-1872) invented the telegraph; in 1876 Alexander Graham Bell (1847-1922) invented the telephone; the incandescent lamp (the light bulb) was invented by Thomas Edison (1847-1931) in 1878; and the first electric motor was invented by Nicholas Tesla (1856-1943) in 1888 (Faraday had built a primitive model of one in 1821). These inventions required the further generation and harnessing of electricity, so efforts were concentrated on developing ways to produce more and more power and to create better equipment, such as motors and transformers.

Edison's invention led to a dependence on electricity for lighting our homes, work areas, and streets. He later created the phonograph and other electrical instruments, which led to the establishment of his General Electric Company. One of today's major telephone companies also had its beginnings during this time. Alexander Bell's invention led to the establishment of the Bell Telephone Company, which eventually became American Telephone and Telegraph (AT&T).

The roots of electronics, which is distinguished from the science of electricity by its focus on lower power generation, can also be found in the 19th century. In the late 1800s, current moving through space was observed for the first time; this was called the "Edison effect." In the early 20th century, devices (such as vacuum tubes, which are pieces of metal inside a glass bulb) were invented that could transmit weak electrical signals, leading to the potential transmission of electromagnetic waves for communication—radio broadcast. The unreliability of vacuum tubes led to the invention of equipment that could pass electricity through solid materials; hence transistors came to be known as solid-state devices.

In the 1960s, transistors were being built on tiny bits of silicon, which became known as microchips. The computer industry is a major beneficiary of the creation of these circuits, because vast amounts of information can be stored on just one tiny chip smaller than a dime.

Knowledge about microchips led to the development of microprocessors. Microprocessors are silicon chips on which the logic and arithmetic functions of a computer are placed. Microprocessors serve as miniature computers and are used in many types of products. The miniaturization of electronic components allowed scientists and engineers to make smaller, lighter computers that could perform the same, or additional, functions of larger computers. They also allowed for the development of many new products. At first they were used primarily in desktop calculators, video games, digital watches, telephones, and microwave ovens. Today, microprocessors are used in electronic controls of automobiles, personal computers, telecommunications systems, and many other products. As a leader in advanced technology, the electronics industry is one of the most important industries today.

The Job

Because electrical and electronics engineering is such a diverse field, there are numerous divisions and departments within which engineers work. In fact, the discipline reaches nearly every other field of applied science and technology. In general, electrical and electronics engineers use their knowledge of the sciences in the practical applications of electrical energy. They concern themselves with things as large as atom smashers and as small as microchips. They are involved in the invention, design, construction, and operation of electrical and electronic systems and devices of all kinds.

The work of electrical and electronics engineers touches almost every niche of our lives. Think of the things around you that have been designed, manufactured, maintained, or in any other way affected by electrical energy: the lights in a room, cars on the road, televisions, stereo systems, telephones, your doctor's blood-pressure reader, computers. When you start to think in these terms, you will discover that the electrical engineer has in some way had a hand in science, industry, commerce, entertainment, and even art.

The list of specialties that engineers are associated with reads like an alphabet of scientific titles—from acoustics, speech, and signal processing; to electromagnetic compatibility; geoscience and remote sensing; lasers and electro-optics; robotics; ultrasonics, ferroelectrics, and frequency control; to vehicular technology. As evident in this selected list, engineers are apt to specialize in what interests them, such as communications, robotics, or automobiles.

As mentioned earlier, electrical engineers focus on high-power generation of electricity and how it is transmitted for use in lighting homes and powering factories. They are also concerned with how equipment is designed and maintained and how communications are transmitted via wire and airwaves. Some are involved in the design and construction of power plants and the manufacture and maintenance of industrial machinery.

Electronics engineers work with smaller-scale applications, such as how computers are wired, how appliances work, or how electrical circuits are used in an endless number of applications. They may specialize in computers, industrial equipment and controls, aerospace equipment, or biomedical equipment.

Tom Busch is an electrical engineer for the U.S. government. He works at the Naval Surface Warfare Center, Crane Division, and much of his work involves testing equipment that will be used on the Navy's ships. "We get equipment that government contractors have put together and test it to make sure it is functioning correctly before it goes out to the fleet," he says. "The type of equipment we test might be anything from navigation to propulsion to communications equipment." Although much of his work currently focuses on testing, Tom also does design work. "We do some software design and also design circuits that go in weapons systems," he says.

Design and testing are only two of several categories in which electrical and electronics engineers may find their niche. Others include research and development, production, field service, sales and marketing, and teaching. In addition, even within each category there are divisions of labor.

Researchers concern themselves mainly with issues that pertain to potential applications. They conduct tests and perform studies to evaluate fundamental problems involving such things as new materials and chemical interactions. Those who work in design and development adapt the researchers' findings to actual practical applications. They devise functioning devices and draw up plans for their efficient production, using computer-aided design and engineering (CAD/CAE) tools. For a typical product such as a television, this phase usually takes up to 18 months to accomplish. For other products, particularly those that utilize developing technology, this phase can take as long as ten years or more.

Production engineers have perhaps the most hands-on tasks in the field. They are responsible for the organization of the actual manufacture of whatever electric product is being made. They take care of materials and machinery, schedule technicians and assembly workers, and make sure that standards are met and products are quality-controlled. These engineers must have access to the best tools for measurement, materials handling, and processing.

After electrical systems are put in place, *field service engineers* must act as the liaison between the manufacturer or distributor and the client. They ensure the correct installation, operation, and maintenance of systems and products for both industry and individuals. In the sales and marketing divisions, engineers stay abreast of customer needs in order to evaluate potential applications, and they advise their companies of orders and effective marketing. A *sales engineer* would contact a client interested in, say, a certain type of microchip for its automobile electrical system controls. He or she would learn about the client's needs and report back to the various engineering teams at his or her company. During the manufacture and distribution of the product, the sales engineer would continue to communicate information between company and client until all objectives were met.

All engineers must be taught their skills, and so it is important that some remain involved in academia. *Professors* usually teach a portion of the basic engineering courses as well as classes in the subjects that they specialize in. Conducting personal research is generally an ongoing task for professors in addition to the supervision of student work and student research. A part of the teacher's time is also devoted to providing career and academic guidance to students.

Whatever type of project an engineer works on, he or she is likely to have a certain amount of desk work. Writing status reports and communicating with clients and others who are working on the same project are examples of the paperwork that most engineers are responsible for. Tom says that the amount of time he spends doing desk work varies from project to project. "Right now, I probably spend about half of my time in the lab and half at my desk," he says. "But it varies, really. Sometimes, I'm hardly in the lab at all; other times, I'm hardly at my desk."

Requirements

High School

Electrical and electronics engineers must have a solid educational background. The discipline is based on much in the applied sciences but requires a clear understanding of practical applications. To prepare for college, high school students should take classes in algebra, trigonometry, calculus, biology, physics, chemistry, computer science, word processing, English, and social studies. According to Tom, business classes are also a good idea. "It wouldn't hurt to get some business understanding—and computer skills are tremendously important for engineers, as well," he says. Students who are planning to pursue studies beyond a bachelor of science degree will also need to take a foreign language. It is recommended that students aim for honors-level courses.

Postsecondary Training

Tom's educational background includes a bachelor of science degree in electrical engineering. Other engineers might receive similar degrees in electronics, computer engineering, or another related science. Numerous colleges and universities offer electrical, electronics, and computer engineering programs. Because the programs vary from one school to another, you should explore as many schools as possible to determine which program is most suited to your academic and personal interests and needs. Most engineering programs have strict admission requirements and require students to have excellent academic records and top scores on national college-entrance examinations. Competition can be fierce for some programs, and high school students are encouraged to apply early.

Many students go on to receive a master of science degree in a specialization of their choice. This usually takes an additional two years of study beyond a bachelor's program. Some students pursue a master's degree immediately upon completion of a bachelor's degree. Other students, however, gain work experience first and then take graduate-level courses on a part-time basis while they are employed. A doctoral degree, or Ph.D., is also available. It generally requires four years of study and research beyond the bachelor's degree and is usually completed by people interested in research or teaching.

By the time you reach college, it is wise to be considering which type of engineering specialty you might be interested in. In addition to the core engineering curriculum (advanced mathematics, physical science, engineering science, mechanical drawing, computer applications), students will begin to choose from the following types of courses: circuits and electronics, signals and systems, digital electronics and computer architecture, electromagnetic waves, systems, and machinery, communications, and statistical mechanics.

Other Requirements

To be a good electrical or electronics engineer, you should have strong problem-solving abilities, mathematical and scientific aptitudes, and the willingness to learn throughout one's career. According to Tom, a curiosity for how things work is also important. "I think you have to like to learn about things," he says. "I also think it helps to be kind of creative, to like to make things."

Most engineers work on teams with other professionals, and the ability to get along with others is essential. In addition, strong communications skills are needed. Engineers need to be able to write reports and give oral presentations.

Exploring

People who are interested in the excitement of electricity can tackle experiments such as building a radio or central processing unit of a computer. Special assignments can also be researched and supervised by teachers. Joining a science club, such as the Junior Engineering Technical Society (JETS), can provide hands-on activities and opportunities to explore scientific topics in depth. Student members can join competitions and design structures that exhibit scientific know-how. Reading trade publications, such as the *JETS Report,* are other ways to learn about the engineering field. This magazine includes articles on engineering-related careers and club activities.

Students can also learn more about electrical and electronics engineering by attending a summer camp or academic program that focuses on scientific projects as well as recreational activities. For example, the Delphian School in Oregon holds summer sessions for high school students. Students are involved in leadership activities and special interests such as computers. Sports and wilderness activities are also offered. Summer programs such as the one offered by the Michigan Technological University focus on career exploration in computers, electronics, and robotics. This academic program for high school students also offers arts guidance, wilderness events, and other recreational activities. (For further information on clubs and programs, write to the sources listed at the end of this article.)

Employers

Most electrical and electronics engineers work in industry, often for design and manufacturing companies or consulting agencies. Others, like Tom, work for the federal government, as teachers in engineering schools and programs, and in research. Some work as private consultants.

Starting Out

Many students begin to research companies that they are interested in working for during their last year of college or even before. It is possible to research companies using many resources, such as company directories and annual reports, available at public libraries. For example, *The Career Guide,*

Dun's Employment Opportunity Directory lists companies that are employing people in electrical/electronics engineering positions, as well as other careers. It gives brief company profiles, describes employment opportunities within the company, and provides addresses to which applicants can write.

Employment opportunities can be found through a variety of sources. Many engineers are recruited by companies while they are still in college. This is what happened to Tom. "I was interviewed while I was still on campus, and I was hired for the job before I graduated," he says. Other companies have internship, work-study, or cooperative education programs from which they hire students who are still in college. Students who have participated in these program often receive permanent job offers through these companies, or they may obtain useful contacts that can lead to a job interview or offer. Some companies use employment agencies and state employment offices. Companies may also advertise positions through advertisements in newspapers and trade publications. In addition, many newsletters and associations post job listings on the Internet.

Interested applicants can also apply directly to a company they are interested in working for. A letter of interest and resume can be sent to the director of engineering or the head of a specific department. One may also apply to the personnel or human resources departments.

Advancement

Engineering careers usually offer many avenues for advancement. An engineer straight out of college will usually take a job as an entry-level engineer and advance to higher positions after acquiring some job experience and technical skills. Engineers with strong technical skills who show leadership ability and good communications skills may move into positions that involve supervising teams of engineers and making sure they are working efficiently. Engineers can advance from these positions to that of a *chief engineer*. The chief engineer usually oversees all projects and has authority over project managers and managing engineers.

Many companies provide structured programs to train new employees and prepare them for advancement. These programs usually rely heavily on formal training opportunities such as in-house development programs and seminars. Some companies also provide special programs through colleges, universities, and outside agencies. Engineers usually advance from junior-level engineering positions to more senior-level positions through a series of positions. Engineers may also specialize in a specific area once they have acquired the necessary experience and skills.

Some engineers move into sales and managerial positions, with some engineers leaving the electronics industry to seek top-level management positions with other type of firms. Other engineers set up their own firms in design or consulting. Engineers can also move into the academic field and become teachers at high schools or universities.

The key to advancing in the electronics field is keeping pace with technological changes, which occur rapidly in this field. Electrical and electronics engineers will need to pursue additional training throughout their careers in order to stay up-to-date on new technologies and techniques.

Earnings

Starting salaries for all engineers are generally much higher than for workers in any other field. In 1997, entry-level electrical and electronics engineers with a bachelor's degree earned an average of $39,500. Electrical and electronics engineers with a master's degree averaged around $45,000 in their first jobs after graduation. A mid-level engineer, with a few years of experience might earn $50,000. Those who rise to the top of their fields can make more than $100,000 annually. The average annual salary for engineers who work for the government is around $62,000.

Most companies offer attractive benefits packages, although the actual benefits vary from company to company. Benefits can include any of the following: paid holidays, paid vacations, personal days, sick leave; medical, health, life insurance; short- and long-term disability insurance; profit sharing; 401(k) plans; retirement and pension plans; educational assistance; leave time for educational purposes; and credit unions. Some companies also offer computer purchase assistance plans and discounts on company products.

Work Environment

Tom's work hours are typically regular—9:00 to 5:00, Monday through Friday—although there is occasional overtime. In many parts of the country, this 5-day, 40-hour work week is still the norm, but it is becoming much less common. Many engineers regularly work 10 or 20 hours of overtime a week. Engineers in research and development, or those conducting experiments, often need to work at night or on weekends. Workers who supervise production activities may need to come in during the evenings or on weekends to handle special production requirements. In addition to the time spent on the job, many engineers also participate in professional associations and pursue additional training during their free time. Many high-tech companies allow flex-time, which means that workers can arrange their own schedules within certain time frames.

Most electrical and electronics engineers work in fairly comfortable environments. Engineers involved in research and design may, like Tom, work in specially equipped laboratories. Engineers involved in development and manufacturing work in offices and may spend part of their time in production facilities. Depending on the type of work one does, there may be extensive travel. Engineers involved in field service and sales spend a significant time traveling to see clients. Engineers working for large corporations may travel to other plants and manufacturing companies, both around the country and at foreign locations.

Engineering professors spend part of their time teaching in classrooms, part of it doing research either in labs or libraries, and some of the time still connected with industry.

Outlook

More engineers work in the electrical and electronics field than in any other division of engineering. In the United States, there were approximately 367,000 such engineers holding jobs in the industry in 1996. Most worked in engineering and business consulting firms, manufacturing companies that produce electrical and electronic equipment, business machines, computers and data processing companies, and telecommunications parts. Others work for companies that make automotive electronics, scientific equipment, and aircraft parts; consulting firms; public utilities; and government agencies.

The demand for electrical and electronics engineers fluctuates with changes in the economy. In the late 1980s and early 1990s, many companies that produced defense products suffered from cutbacks in defense orders and, as a result, made reductions in their engineering staffs. However, opportunities in defense-related fields may improve, as there is a growing trend toward upgrading existing aircraft and weapons systems. In addition, the increased use of electronic components in automobiles and increases in computer and telecommunications production require a high number of skilled engineers. Opportunities for electrical and electronics engineers are expected to increase faster than the average for all other jobs through 2006.

The growing consumer, business, and government demand for improved computers and communications equipment is expected to propel much of this expected growth. Another area of high demand is projected to be the development of electrical and electronic goods for the consumer market. The strongest job growth, however, is likely to be in nonmanufacturing industries. This is because more and more firms are contracting for electronic engineering services from consulting and service firms.

Engineers will need to stay on top of changes within the electronics industry and will need additional training throughout their careers to learn new technologies. Economic trends and conditions within the global marketplace have become increasingly more important. In the past, most electronics production was done in the United States or by American-owned companies. During the 1990s, this changed, and the electronics industry entered an era of global production. Worldwide economies and production trends will have a larger impact on U.S. production, and companies that cannot compete technologically may not succeed. Job security is no longer a sure thing, and many engineers can expect to make significant changes in their careers at least once. Engineers who have a strong academic foundation, who have acquired technical knowledge and skills, and who stay up-to-date on changing technologies provide themselves with the versatility and flexibility to succeed within the electronics industry.

For More Information

For information on careers and educational programs, please contact the following associations:

Institute of Electrical and Electronics Engineers (IEEE)
1828 L Street, NW, Suite 1202
Washington, DC 20036-5104
Tel: 202-785-0017
Web: http://www.ieee.org/

Electronic Industries Alliance
2500 Wilson Boulevard
Arlington, VA 22201-3834
Tel: 703-907-7500
Web: http://www.eia.org

For information on careers, educational programs, and student clubs, please contact:

Junior Engineering Technical Society, Inc.
1420 King Street, Suite 405
Alexandria, VA 22314-2794
Tel: 703-548-5387
Web: http://www.jets.org

For information on the Summer at Delphi Youth Program for high school students, please contact:

The Delphian School
20950 SW Rock Creek Road
Sheridan, OR 97378
Tel: 800-626-6610
Web: http://www.delphian.org

For information on its summer youth program for high school students, please contact:

Michigan Technological University Summer Youth Program
Youth Programs Office
1400 Townsend Drive
Houghton, MI 49931-1295
Tel: 906-487-1885
Web: http://www.mtu.edu

Environmental Engineers

Overview

Environmental engineers design, build, and maintain systems to control waste streams produced by municipalities or private industry. Such waste streams may be wastewater, solid waste, hazardous waste, or contaminated emissions to the atmosphere (air pollution). Environmental engineers typically are employed by the Environmental Protection Agency (EPA), by private industry, or by engineering consulting firms.

History

The job of environmental engineer—like environmental lawyer, environmental planner, environmental quality analyst, and many others—is an excellent example of a professional category that has evolved to meet the booming needs of the environmental industry. Although people have been doing work that falls into the category of environmental engineering for decades, it is only within about the last 30 years that a separate professional category has been recognized for environmental engineers.

"In the 1930s, 1940s, 1950s, even the 1960s, someone who wanted to be an environmental engineer would have been steered toward sanitary engineering, which basically deals with things like wastewater, putting sewers down," says Lee DeAngelis, regional director of the Environmental Careers Organization (ECO).

Sanitary engineering, in turn, is a form of civil engineering. "Civil engineering is engineering for municipalities," explains Mike Waxman, who heads the environmental training arm of the outreach department at the University of Wisconsin-Madison College of Engineering. "It includes things like building roads, highways, buildings. But a big part of civil engineering is dealing with the waste streams that come from cities or municipalities. Wastewater from a city's sewage treatment plants is a prime example," Mike says. This water must be treated in order to be pure enough to be used again. "Scientists work out what must be done to break down the harmful substances in the water, such as by adding bacteria; engineers design, build, and maintain the systems needed to carry this out. Technicians monitor the systems, take samples, run tests, and otherwise ensure that the system is working as it should."

This structure—scientists deciding what should be done at the molecular or biological level, engineers designing the systems needed to carry out the project, and technicians taking care of the day-to-day monitoring of the systems—is applied to other waste streams as well, Mike adds.

Environmental engineering, then, is an offshoot of civil engineering/sanitary engineering and focuses on the development of the physical systems needed to control waste streams. Civil engineers who already were doing this type of work began to refer to themselves as environmental engineers around 1970, with the great boom in new environmental regulations, according to Mike. "It's what they wanted to be called," he says. "They wanted the recognition for what they were doing."

The Job

Let's say there's a small pond in Crawford County, Illinois. Normally, several different kinds of fish swim lazily through its waters, frogs hop around its banks, and birds build their nests in nearby trees. About a half-mile away is the Jack J. Ryan and Sons Manufacturing Company. For years, this plant has safely treated its wastewater—the effluent produced during the manufacturing process-and discharged it into the pond. Then one day, without warning, hundreds of dead fish wash up on the banks of the pond. What's going on? What should be done? What you would do as an environmental engineer depends on several factors, including your employer and the area in which you specialize.

If you worked for the federal or state Environmental Protection Agency (EPA), your role as an environmental engineer would be like that of a police officer or detective. Your knowledge of systems to treat waste streams—coupled with your authority to enforce environmental regulations and your intimate knowledge of your territory—would make you well qualified to get to the bottom of problems stemming from systems that aren't functioning properly.

How would you tackle the Crawford County pond problem? Let's say you work in the Champaign regional office of the Illinois Environmental Protection Agency (IEPA). There are three divisions—air, land, and water—and you work in water. Your territory includes Crawford County. Alerted to the fish kill at the pond, you get into one of the state vehicles parked outside your office and head out to the site to investigate.

Once there, you snap pictures, take samples of the water, make notes. You're documenting the problem, but you're also asking yourself questions based on your knowledge of the area: Is it a discharge problem from Jack J. Ryan and Sons? If so, was there an upset in the process? A spill? A flood? Could a storage tank be leaking? Or is the problem further upstream? You know the pond is connected to other waterways; could some other discharger be responsible for killing the fish?

You'd probably pay a visit to Jack J. Ryan and Sons next. You'd talk to people there, such as the production manager. You might ask him if they have been doing anything differently lately. You also might look at plans of the plant. If the mystery remains, you might look further upstream, checking to see if other manufacturers or other wastewater dischargers are doing something new that's caused the fish kill.

When you locate the problem, the next step is enforcement. Let's say the production manager at Jack J. Ryan and Sons tells you yes, they've changed something in the manufacturing process. They're producing a new kind of die-cast part. They're sorry; they didn't know they were doing something wrong. They'll get on the problem right away.

At this point, sensing cooperation, you back off. You tell them they'll be fined $10,000, and you'll be checking back with them soon to see what they've done. But you won't turn the case over to the lawyers—yet.

Let's say that instead of working for the EPA, you're on the other side of the problem: you work for Jack J. Ryan and Sons' environmental staff. Your job is to help get the company into compliance and keep it that way, all the while balancing the economic concerns of your employer. (At one time, industries' environmental affairs positions may have been filled by workers from the plant. Since the late 1980s, however, they tend to be staffed by people dedicated strictly to environmental matters, including scientists, engineers, lawyers, and communications professionals.) One day, you get a call from an engineer at the IEPA: "There seems to be a fish kill at the pond near your plant. We've determined it's probably from a discharge from your plant." At this point, you'd jump up and get busy. You'd probably look at your plant's plans, talk to the production manager, and figure out a plan of action. If you were having trouble coming up with a plan, you might turn to a consulting engineering firm for help.

As an environmental engineer with a consulting firm, you'd apply your expertise to problems your firm's clients were having. If Jack J. Ryan and Sons called your company for help, for example, you might be part of a team that goes out to the plant, assesses the problem, and designs a system to get the plant back into compliance. Consulting firms balance what the client wants, needs, and can afford. They know the technical aspects of waste control and also may sell clients on their expertise in dealing with the government—filling out the required government forms, for example.

Broadly speaking, environmental engineers may focus on one of three areas: *air, land,* or *water.* Air includes air pollution control, air quality management, and other specialties involved in dealing with systems to treat emissions. The private sector tends to have the majority of these jobs, according to ECO. Land includes landfill professionals, for whom environmental engineering and public health are key areas. Water includes activities like those described above.

A big area for environmental engineers is hazardous waste management. Expertise in designing systems and processes to reduce, recycle, and treat hazardous waste streams is very much in demand, according to ECO. This area tends to be the most technical of all the environmental fields and so demands more professionals with graduate and technical degrees.

Environmental engineers spend a lot of time on paperwork—including writing reports and memos and filling out forms. They also might climb a smokestack, wade in a creek, or go toe-to-toe with a district attorney in a battle over a compliance matter. If they work on in-house staffs, they may face frustration over not knowing what is going on in their own plants. If they work for the government, they might struggle with bureaucracy. If they work for a consultant, they may have to juggle the needs of the client (including the need to keep costs down) with the demands of the government.

Requirements

High School

A bachelor's degree is mandatory to work in environmental engineering. At the high school level, the most important course work is in science and mathematics. It's also good to develop written communication skills. Competition to get into the top engineering schools is tough, so make sure you do well on your ACT or SAT tests.

Postsecondary Training

At this writing, about 20 schools offer an undergraduate degree in environmental engineering. Other possibilities are to earn a civil engineering, mechanical engineering, industrial engineering, or other traditional engineering degree with an environmental focus, to obtain a traditional engineering degree and pick up the environmental knowledge on the job, or to obtain a masters' degree in environmental engineering.

Certification or Licensing

If your work as an engineer affects public health, safety, or property, you must register with the state. To obtain registration, you must have a degree from an accredited engineering program. Right before you get your degree (or soon after), you must pass an engineer-in-training (EIT) exam covering fundamentals of science and engineering. A few years after you've started your career, you also must pass an exam covering engineering practice. Additional certification is voluntary and may be obtained through such organizations as the American Academy of Environmental Engineers.

Other Requirements

People who like solving problems, have a good background in science and math—who could draw on their knowledge of differential equations if they had to, for example—and could, in the words of one engineer, "just get in there and figure out what needs to be done," are good candidates for this position.

Exploring

A good way to explore becoming an environmental engineer is to talk to someone in the field. Contact your local EPA office, check the Yellow Pages for environmental consulting firms in your area, or ask a local industrial company if you can visit. The latter is not as far-fetched as you might think: big industry has learned the value of earning positive community relations, and their outreach efforts may include having an open house for their neighbors in which one can walk through their plants, ask questions, and get a feel for what goes on there.

You cannot practice at being an environmental engineer without having a bachelor's degree. However, you can put yourself in situations in which you're around environmental engineers to see what they do and how they work. To do so, you may volunteer for the local chapter of a nonprofit environmental organization, do an internship through ECO or another organization, or work first as an environmental technician, a job that requires less education (such as a two-year associate's degree or even a high school diploma).

Another good way to get a feel for a field is to familiarize yourself with its professional journals. Two journals that may be available in your library include *Chemical & Engineering News,* which regularly features articles on waste management systems, and *Pollution Engineering,* which features articles of interest to environmental engineers.

Employers

Environmental engineers most often work for the Environmental Protection Agency (EPA), in private industry, or at engineering consulting firms.

Starting Out

The traditional method of entering this field is by obtaining a bachelor's degree and applying directly to companies or to the EPA. School placement offices can assist you in these efforts.

Advancement

After you've worked for a time as an environmental engineer, there are several routes for advancement. If you start out with the EPA, you may become a department supervisor or switch to private industry or consulting. In-house environmental staff members may rise to supervisory positions. Engineers with consulting firms may become project managers or specialists in certain areas.

Environmental careers are evolving at a breakneck speed. New specialties are emerging all the time. Your advancement may take the form of getting in on—maybe even helping to develop—some subspecialty yet to be invented that suits your own particular interests, experience, and expertise.

Earnings

The following salary information is from ECO:

Average solid waste management pay is slightly lower than that for hazardous waste management. Entry-level salaries for professionals in this field range from less than $20,000 to $30,000, with engineers at the higher end of the scale. In water quality management, the range is about $30,000 to $40,000 for state and federal government jobs and $30,000 and up for private jobs.

Fringe benefits vary widely depending on the employer. State EPA jobs may include, for example, two weeks of vacation, health insurance, tuition reimbursement, use of company vehicles for work, and similar perks. In-house or consulting positions may add additional benefits to lure top candidates.

Work Environment

You're likely to split your time between working in an office and working out in the field (or, if you're a member of an in-house staff, out in your plant or at the site of discharges). You also may go to court. Since ongoing education is crucial in most of these positions, you'll spend time in school, at workshops, and studying up on new regulations, techniques, and problems. It's important to be able to work as part of a team that may include any of a number of different specialists. It's also important to be able to communicate well, both in writing and in discussions. You must be comfortable with technical information, be able to apply your background in math and science, and preferably be able to communicate often-complex engineering information to people who may not have the type of technical background that you do.

Outlook

The 1980s were a time of increased environmental regulation and enforcement. Superfund legislation forced states to clean up hazardous waste sites and the U.S. Environmental Protection Agency required companies to reduce waste and dispose of it more responsibly. Environmental engineers, consequently, had abundant opportunities. In the 1990s, many of the major cleanup efforts were undertaken or finished, causing the environmental engineering job market to taper off from its rapid growth.

The Clean Air Act of 1990 did create a brief surge in air pollution jobs. Overall, however, the water supply and water pollution control specialties presently offer the most job opportunities for environmental engineers.

Opportunities are available with all three major employers—the EPA, industry, and consulting firms. The EPA has long been a big employer of environmental engineers. At this writing, private industry and consultants are hiring more and more of them.

For More Information

For information on certification, careers, and salaries or a copy of Environmental Engineering Selection Guide *(giving names of accredited environmental engineering programs and of professors who have board certification as environmental engineers), contact:*

American Academy of Environmental Engineers
130 Holiday Court, Suite 100
Annapolis, MD 21401
Tel: 410-266-3311
Web: http://www.enviro-engrs.org

A cross-disciplinary environmental association:

National Association of Environmental Professionals
6524 Ramoth Drive
Jacksonville, FL 32226-3202
Tel: 904-251-9900
Web: http://www.naep.org

National Solid Wastes Management Association
4301 Connecticut Avenue, NW, Suite 300
Washington, DC 20008
Tel: 202- 244-4700
Web: http://www.envasns.org/nswma

Contact SCA for information about internships for high school students:

Student Conservation Association
689 River Road
PO Box 550
Charlestown, NH 03603-0550
Tel: 603-543-1700
Web: http://www.sca-inc.org

Industrial Engineers

Computer science Mathematics	School Subjects
Leadership/management Technical/scientific	Personal Skills
Primarily indoors Primarily one location	Work Environment
Bachelor's degree	Minimum Education Level
$38,000 to $52,300 to $90,000	Salary Range
Required by certain states	Certification or Licensing
About as fast as the average	Outlook

Overview

Industrial engineers use their knowledge of various disciplines—including systems engineering, management science, operations research, and fields such as ergonomics—to determine the most efficient and cost-effective methods for industrial production. Engineers are responsible for designing systems that integrate materials, equipment, information, and people in the overall production process.

History

In today's industries, manufacturers increasingly depend on industrial engineers to determine the most efficient production techniques and processes. The roots of industrial engineering, however, can be traced to ancient Greece, where records indicate that manufacturing labor was divided among people having specialized skills.

The most significant milestones in industrial engineering—before the field even had an official name—occurred in the 18th century, when a number of inventions were introduced in the textile industry. The first was the flying shuttle that opened the door to the highly automatic weaving we now take for granted. This shuttle allowed one person, rather than two, to weave fabrics wider than ever before. Other innovative devices, such as the power loom and the spinning jenny that increased weaving speed and improved quality, soon followed. By the late 18th century, the Industrial Revolution was in full swing. Innovations in manufacturing were made, standardization of interchangeable parts was implemented, and specialization of labor was increasingly put into practice.

Industrial engineering as a science is said to have originated with the work of Frederick Taylor. In 1881, he began to study the way production workers used their time. At the Midvale Steel Company where he was employed, he introduced the concept of *time study,* whereby workers were timed with a stopwatch and their production was evaluated. He used the studies to design methods and equipment that allowed tasks to be done more efficiently.

In the early 1900s, the field was known as *scientific management.* Frank and Lillian Gilbreth were influential with their motion studies of workers performing various tasks. Then, around 1913, automaker Henry Ford implemented a conveyor belt assembly line in his factory, which led to increasingly integrated production lines in more and more companies. Industrial engineers nowadays are called upon to solve ever more complex operating problems and to design systems involving large numbers of workers, complicated equipment, and vast amounts of information. They meet this challenge by utilizing advanced computers and software to design complex mathematical models and other simulations.

The Job

Industrial designers are involved with the development and implementation of the systems and procedures that are utilized by many industries and businesses. In general, they figure out the most effective ways to use the three basic elements of any company: people, facilities, and equipment.

Although industrial engineers work in a variety of businesses, the main focus of the discipline is in manufacturing, also called industrial production. Primarily, industrial engineers are concerned with process technology, which includes the design and layout of machinery and the organization of workers who implement the required tasks.

Industrial engineers' responsibilities are numerous. With regard to facilities and equipment, engineers are involved in selecting machinery and other equipment and then in setting them up in the most efficient production layout. They also develop methods to accomplish production tasks, such as the organization of an assembly line. In addition, they devise systems for quality control, distribution, and inventory.

Industrial engineers are responsible for some organizational issues. For instance, they might study an organization chart and other information about a project and then determine the functions and responsibilities of workers. They devise and implement job evaluation procedures as well as articulate labor-utilization standards for workers. Engineers often meet with managers to discuss cost analysis, financial planning, job evaluation, and salary administration. Not only do they recommend methods for improving employee efficiency but they may also devise wage and incentive programs.

Industrial engineers evaluate ergonomic issues—the relationship between human capabilities and the physical environment in which they work. For example, they might evaluate whether machines are causing physical harm or discomfort to workers, or whether the machines could be designed differently to enable workers to be more productive.

In industries that do not focus on manufacturing, industrial engineers are often called *management analysts* or *management engineers*. In the health care industry, such engineers are asked to evaluate current administrative and other procedures. They also advise on job standards, cost-containment, and operations consolidation. Some industrial engineers are employed by financial services companies. Because many engineering concepts are relevant in the banking industry, engineers there design methods to optimize the ratio of tellers to customers and the use of computers for various tasks and to handle mass distribution of items such as credit card statements.

Requirements

High School

To prepare for a college engineering program, concentrate on mathematics (algebra, trigonometry, geometry, calculus), physical sciences (physics, chemistry), social sciences (economics, sociology), and English. Engineers often have to convey ideas graphically and may need to visualize processes in three-dimension, so courses in graphics, drafting, or design are also help-ful. Students should take honors level courses if possible.

Postsecondary Training

A bachelor's degree from an accredited institution is usually the minimum requirement for all professional positions. There are about 100 accredited industrial engineering programs in the United States. Colleges offer either four- or five-year engineering programs. Because of the intensity of the curricula, many students take heavy course loads and attend summer sessions in order to finish in four years.

During their junior and senior years, students should be considering specific career goals, such as in which industry to work. Third- and fourth-year courses focus on such subjects as facility planning and design, work measurement standards, process design, engineering economics, manufacturing and automation, and incentive plans.

Many industrial engineers go on for a postgraduate degree. These programs tend to involve more research and independent study. Graduate degrees are usually required for teaching positions.

Certification or Licensing

Registration as a professional engineer is generally voluntary but is often considered when employers are reviewing workers for promotion. Registration guidelines are different in each state. Normally they involve meeting certain educational requirements and passing an eight-hour Fundamentals of Engineering (FE) exam in your junior year. After working a specified amount of time in engineering, you must pass the eight-hour Professional Engineering (PE) exam. Another credential is the Systems Integration Certificate, which is offered by the Institute of Industrial Engineers to those who have been working in the field for at least five years.

Other Requirements

Industrial engineers enjoy problem solving and analyzing things as well as being a team member. The ability to communicate is vital since engineers interact with all levels of management and workers. Being organized and detail-minded is important because industrial engineers often handle large projects and must bring them in on time and on budget. Since process design is the cornerstone of the field, an engineer should be creative and inventive.

Exploring

Try joining a science or engineering club, such as the Junior Engineering Technical Society (JETS). JETS offers academic competitions in subjects such as computer fundamentals, mathematics, physics, and English. It also conducts design contests in which students learn and apply science and engineering principles. Membership in JETS includes a quarterly magazine, *JETS Report*, that has interviews and articles on various engineering careers. You also might read some engineering books for background on the field or magazines such as *Industrial Engineering*.

Other opportunities for exploring industrial engineering careers can be found at summer camps. For example, the Worcester Polytechnic Institute in Massachusetts offers a 12-day session for high school seniors that focuses on programs in science, math, and various engineering disciplines while offering recreational activities.

Employers

Although a majority of industrial engineers are employed in the manufacturing industry, related jobs are found in almost all businesses, including transportation; communications; electric; gas and sanitary services; government; finance; insurance; real estate; wholesale and retail trade; construction; mining; agriculture; forestry; and fishing. Also, many work as independent consultants.

Starting Out

The main qualification for an entry-level job is a bachelor's degree in industrial engineering. Accredited college programs generally have job openings listed in their placement offices. Entry-level industrial engineers find jobs in various departments, such as computer operations, warehousing, and quality control. As engineers gain on-the-job experience and familiarity with departments, they may decide on a specialty. Some may want to continue to work as process designers or methods engineers, while others may move on to administrative positions.

Some further examples of specialties include work measurement standards; shipping and receiving; cost control; engineering economics; materials handling; management information systems; mathematical models; and operations. Many who choose industrial engineering as a career find its appeal in the diversity of sectors that are available to explore.

Advancement

After having worked at least three years in the same job, an industrial engineer may have the basic credentials needed for advancement to a higher position. In general, positions in operations and administration are considered high-level jobs, although this varies from company to company. Engineers who work in these areas tend to earn larger salaries than those who work in warehousing or cost control, for example. If one is interested in moving to a different company, it is considered easier to do so within the same industry.

Industrial engineering jobs are often considered stepping stones to management positions—even in other fields. Engineers with many years' experience frequently are promoted to higher level jobs with greater responsibilities. Because of the field's broad exposure, industrial engineering employees are generally considered better prepared for executive roles than are other types of engineers.

Earnings

According to the U.S. Bureau of Labor Statistics, the average annual wage for industrial engineers in 1997 was $52,350. As with most occupations, salaries rise as more experience is gained. Veteran engineers can earn over $90,000. According to a survey by the National Association of Colleges and Employers, the average starting salary for industrial engineers is $38,000.

Salaries for industrial engineers also vary with regard to geographic location. Those who work in states along the Pacific, for example, tend to make more than those in any other area in the United States. Industrial engineers working in the Great Lakes area, such as in Illinois and Indiana, generally make less than those in the Pacific states but more than those in the north-central part of the country, where salaries in this field are usually lowest.

Work Environment

Industrial engineers usually work in offices at desks and computers, designing and evaluating plans, statistics, and other documents. Overall, industrial engineering is ranked above other engineering disciplines for factors such as employment outlook, salary, and physical environment. However, industrial engineering jobs are considered stressful because they often entail tight deadlines and demanding quotas, and jobs are moderately competitive. Engineers work an average of 46 hours per week.

Industrial engineers generally collaborate with other employees, conferring on designs and procedures, as well as with business managers and consultants. Although they spend most of their time in their offices, they frequently must evaluate conditions at factories and plants, where noise levels are often high.

Outlook

In 1997, industrial engineers held about 112,400 jobs in the United States. The U.S. Bureau of Labor Statistics anticipates that industrial engineering opportunities will grow through the year 2006, however, this increase will not be any faster than the average growth for all occupations.

Engineers who transfer or retire will create the highest percentage of openings in this field. New jobs will be found in growing industries, especially those implementing automation to solve complicated business operations. The demand for industrial engineers will continue as manufacturing and other companies strive to make their production processes more effective and competitive.

Although approximately 75 percent of industrial engineering jobs are in the manufacturing industry, opportunities are found in a diversity of fields because the skills involved can be used in nearly any type of business.

For More Information

For a list of ABET-accredited engineering schools, contact:

Accreditation Board for Engineering and Technology, Inc.
111 Market Place, Suite 1050
Baltimore, MD 21202-4012
Tel: 410-347-7700
Web: http://www.abet.ba.md.us http://www.abet.ba.md.us

For information about colleges and careers in industrial engineering, contact:

Institute of Industrial Engineers
25 Technology Park
Norcross, GA 30092
Tel: 770-449-0461
Web: http://www.iienet.org

Mechanical Engineers

Computer science English Mathematics	School Subjects
Leadership/management Technical/scientific	Personal Skills
Primarily indoors One location with some travel	Work Environment
Bachelor's degree	Minimum Education Level
$34,000 to $52,000 to $85,000	Salary Range
Voluntary	Certification or Licensing
About as fast as the average	Outlook

Overview

Mechanical engineers plan and design tools, engines, machines, and other mechanical systems that produce, transmit, or use power. Their work varies by industry and function. They may work in design, instrumentation, testing, robotics, transportation, or bioengineering, among other areas. The broadest of all engineering disciplines, it extends across many interdependent specialties. Mechanical engineers may work in production operations, maintenance, or technical sales, and many are administrators or managers.

History

In a general sense, mechanical engineering has existed for thousands of years. Pyramid building in ancient Egypt, for example, required extensive knowledge of engineering principles. Large blocks of two- and three-ton stone were quarried, transported, and positioned according to sophisticated designs.

Ancient Greeks and Romans were also great builders, but, unlike the Egyptians, they developed and made use of elaborate mechanical devices, including water pumps, machines for cutting screws, and treadmills that produced power for lifting heavy objects. Remarkably, the Greeks even invented a steam engine, but viewed it only as a curiosity or toy.

The term "engineer" was coined around the 14th century and applied to people who designed equipment for war. Their achievements were so important that the strength of a country's military became increasingly dependent upon their inventions. As these individuals applied their knowledge to civilian needs, new occupational terms developed. Engineers who worked on civilian projects came to be known as civil, as opposed to military, engineers. Later, engineers who concentrated on machinery and the generation of power were called mechanical engineers.

The modern field of mechanical engineering took root during the Renaissance. In this period, engineers focused their energies on developing more efficient ways to perform such ordinary tasks as grinding grain and pumping water. Water wheels and windmills were common energy producers at that time. Leonardo da Vinci, who attempted to design such complex machines as a submarine and a helicopter, best personified the burgeoning mechanical inventiveness of the period. One of the Renaissance's most significant inventions was the mechanical clock, powered first by falling weights and later by compressed springs.

Despite these developments, it was not until the Industrial Revolution that mechanical engineering took on its modern form. The steam engine, an efficient power producer, was introduced in 1712 by Thomas Newcomen to pump water from English mines. More than a half century later, James Watt modified Newcomen's engine to power industrial machines. In 1876, German Nicolaus Otto developed the internal combustion engine, which became one of the century's most important inventions. In 1847, a group of British engineers, who specialized in steam engines and machine tools, organized the Institution of Mechanical Engineers. The American Society of Mechanical Engineers was formed by 1880.

Mechanical engineering has rapidly expanded in the 20th century. More than 200,000 mechanical engineers are employed in the United States alone. Mass production systems allow large quantities of standardized goods to be made at a low cost, and mechanical engineers play a pivotal role in the design of these systems. In the second half of the 20th century, computers revolutionized production. Mechanical engineers now design mechanical systems on computers, and they are used to test, monitor, and analyze mechanical systems and factory production. Mechanical engineers realize this trend is here to stay.

The Job

The work of mechanical engineering begins with research and development. A company may need to develop a more fuel-efficient automobile engine, for example, or a cooling system for air-conditioning and refrigeration that does not harm the earth's atmosphere. A *research engineer* explores the project's theoretical, mechanical, and material problems. The engineer may perform experiments to gather necessary data and acquire new knowledge. Often, an experimental device or system is developed.

The *design engineer* takes information gained from research and development and uses it to plan a commercially useful product. To prevent rotting in a grain storage system, for example, a design engineer might use research on a new method of circulating air through grain. The engineer would be responsible for specifying every detail of the machine or mechanical system. Since the introduction of sophisticated software programs, mechanical engineers have increasingly used computers in the design process.

After the product has been designed and a prototype developed, the product is analyzed by *testing engineers*. A tractor transmission, for example, would need to be tested for temperature, vibration, dust, and performance under the required loads, as well as for any government safety regulations. If dust is penetrating a bearing, the testing engineer would refer the problem to the design engineer, who would then make an adjustment to the design of the transmission. Design and testing engineers continue to work together until the product meets the necessary criteria.

Once the final design is set, it is the job of the *manufacturing engineer* to come up with the most time- and cost-efficient way of making the product, without sacrificing quality. The amount of factory floor space, the type of manufacturing equipment and machinery, and the cost of labor and materials are some of the factors that must be considered. Engineers select the necessary equipment and machines and oversee their arrangement and safe operation. Other engineering specialists, such as chemical, electrical, and industrial engineers, may provide assistance.

Some types of mechanical systems—from machinery on a factory floor to a nuclear power plant—are so sophisticated that mechanical engineers are needed for operation and ongoing maintenance. With the help of computers, *maintenance and operations engineers* use their specialized knowledge to monitor these complex systems and to make necessary adjustments.

Mechanical engineers also work in marketing, sales, and administration. Because of their training in mechanical engineering, *sales engineers* can give customers a detailed explanation of how a machine or system works. They may also be able to alter its design to meet a customer's needs.

In a small company, a mechanical engineer may need to perform many, if not most, of the above responsibilities. Some tasks might be assigned to *consulting engineers,* who are either self-employed or work for a consulting firm.

Other mechanical engineers may work in a number of specialized areas. *Energy specialists* work with power production machines to supply clean and efficient energy to individuals and industries. *Application engineers* specialize in computer-aided design (CAD) systems. *Process engineers* work in environmental sciences to reduce air pollution levels without sacrificing essential services such as those provided by power stations or utility companies.

Requirements

High School

If you are interested in mechanical engineering as a career, you need to take courses in geometry, trigonometry, and calculus. Physics and chemistry courses are also recommended, as is mechanical drawing or computer-aided design, if they are offered at your high school. Communications skills are important for mechanical engineers because they interact with a variety of co-workers and vendors and are often required to prepare and/or present reports. English and speech classes would be helpful. Finally, because computers are such an important part of engineering, computer science courses are good choices.

Postsecondary Training

A bachelor's degree in mechanical engineering is usually the minimum educational requirement for entering this field. A master's degree, or even a Ph.D., may be necessary for obtaining some positions, such as those in research, teaching, and administration.

In the United States, there are more than 200 colleges and universities where engineering programs have been approved by the Accreditation Board for Engineering and Technology (ABET). Most of these institutions offer programs in mechanical engineering. Although admissions requirements vary slightly from school to school, most require a solid background in mathematics and science.

In a four-year undergraduate program, students typically begin by studying mathematics and science subjects, such as calculus, differential equations, physics, and chemistry. Course work in liberal arts and elementary mechanical engineering is also taken. By the third year, students begin to study the technical core subjects of mechanical engineering—mechanics, thermodynamics, fluid mechanics, design manufacturing, and heat transfer—as well as such specialized topics as power generation and transmission, CAD, and the properties of materials.

At some schools, there is a five- or six-year program that combines classroom study with practical experience working for an engineering firm. Although these cooperative, or work-study, programs take longer, they offer significant advantages. Not only does the salary help pay for educational expenses, but the student has the opportunity to apply theoretical knowledge to actual work problems in mechanical engineering. In some cases, the company may offer full-time employment to its co-op workers after graduation.

A graduate degree is a prerequisite for becoming a university professor or researcher. It may also lead to a higher-level job within an engineering department or firm. Some companies encourage their employees to pursue graduate education by offering tuition-reimbursement programs. Because technology is rapidly developing, mechanical engineers need to continue their education, formally or informally, throughout their careers. Conferences, seminars, and professional journals serve to educate engineers about developments in the field.

Certification or Licensing

Engineers whose work may affect the life, health, or safety of the public must be registered according to regulations in all 50 states and the District of Columbia. Applicants for registration must have received a degree from an ABET-accredited engineering program and have four years of experience. They must also pass a written examination.

Many mechanical engineers also become certified. Certification is a status conferred by a professional or technical society for the purpose of recognizing and documenting an individual's abilities in a specific engineering field.

Other Requirements

Personal qualities essential for mechanical engineers include the ability to think analytically, to solve problems, and to work with abstract ideas. Attention to detail is also important, as are good oral and written communications skills and the ability to work well in groups. Computer literacy is essential.

Exploring

One of the best ways to learn about the field is to talk with a mechanical engineer. It might also be helpful to tour an industrial plant or visit a local museum specializing in science and industry. Public libraries usually have books on mechanical engineering that might be enlightening. You might tackle a design or building project to test your aptitude for the field. Finally, some high schools offer engineering clubs or organizations. Membership in the Junior Engineering Technical Society (JETS), a national organization, is suggested for prospective mechanical engineers.

Employers

Most mechanical engineers work in manufacturing, employed by a wide variety of industries. For example, manufacturers of industrial and office machinery, farm equipment, automobiles, petroleum, pharmaceuticals, fabricated metal products, pulp and paper, electronics, utilities, computers, soap and cosmetics, and heating, ventilating, and air-conditioning systems all employ mechanical engineers. Others are self-employed or work for consulting firms, government agencies, or colleges and universities.

Starting Out

Many mechanical engineers find their first job through their college or university placement office. Many companies send recruiters to college campuses to interview and sign up engineering graduates. Other students might find a position in the company where they had a summer or part-time job. Newspapers and professional journals often list job openings for engineers. Job seekers who wish to work for the federal government should contact the nearest branch of the Office of Personnel Management.

Advancement

As engineers gain experience, they can advance to jobs with a wider scope of responsibility and higher pay. Some of these higher level jobs include technical service and development officers, team leaders, research directors, and managers. Some mechanical engineers use their technical knowledge in sales and marketing positions, while others form their own engineering business or consulting firm.

Many engineers advance by furthering their education. A master's degree in business administration, in addition to an engineering degree, is sometimes helpful in obtaining an administrative position. A master's or doctoral degree in an engineering specialty may also lead to executive work. In addition, those with graduate degrees often have the option of research or teaching positions.

Earnings

Starting salaries for mechanical engineers average around $34,000 a year, according to the U.S. Bureau of Labor Statistics. Typically, compensation is considerably higher for those with a graduate degree or more experience. Mid-level engineers earn about $52,000. Those with a great deal of experience can earn annual salaries of $85,000 or more.

Like most professionals, mechanical engineers who work for a company usually receive a generous benefits package, including vacation days, sick leave, health and life insurance, and a savings and pension program. Self-employed mechanical engineers must provide their own benefits.

Work Environment

The working conditions of mechanical engineers vary. Most work indoors in offices, research laboratories, or production departments of factories and shops. Depending on the job, however, a significant amount of work time may be spent on a noisy factory floor, at a construction site, or at another field operation. Mechanical engineers have traditionally designed systems on drafting boards, but, since the introduction of sophisticated software programs, design is increasingly done on computers.

Engineering is, for the most part, a cooperative effort. While the specific duties of an engineer may require independent work, each project is typically the job of an engineering team. Such a team might include other engineers, engineering technicians, and engineering technologists.

Mechanical engineers generally have a 40-hour workweek; however, their working hours are often dictated by project deadlines. They may work long hours to meet a deadline, or show up on a second or third shift to check production at a factory or a construction project.

Mechanical engineering can be a very satisfying occupation. Engineers often get the pleasure of seeing their designs or modifications put into actual, tangible form. Conversely, it can be frustrating when a project is stalled, full of errors, or even abandoned completely.

Outlook

More than three out of five mechanical engineers work in manufacturing fields, such as machinery, transportation (including automotive) equipment, electrical equipment, and fabricated metal products. The remainder are largely found in government agencies and consulting firms.

The employment of mechanical engineers is expected to grow about as fast as average through the year 2006, according the Bureau of Labor Statistics. Although overall employment in manufacturing is expected to decline, engineers will be needed to meet the demand for more efficient industrial machinery and machine tools. Employment for mechanical engineers in business and engineering services firms is expected to grow faster than the average, as other industries increasingly contract out to these firms to solve engineering problems. It should also be noted that reductions in defense spending may adversely affect the employment outlook for engineers within the federal government.

For More Information

For information a list of engineering programs at colleges and universities, contact:

Accreditation Board for Engineering and Technology, Inc.
111 Market Place, Suite 1050
Baltimore, MD 21202
Tel: 410-347-7700
Web: http://www.abet.ba.md.us

For information on mechanical engineering and mechanical engineering technology, contact:

American Society of Mechanical Engineers
Three Park Avenue
New York, NY 10016
Tel: 212-591-7000 or 800-THE-ASME
Web: http://www.asme.org

For information about careers and high school engineering competitions, contact:

Junior Engineering Technical Society, Inc.
1420 King Street, Suite 405
Alexandria, VA 22314
Tel: 703-548-5387
Web: http://www.jets.org

Metallurgical Engineers

Chemistry Physics	School Subjects
Leadership/management Technical/scientific	Personal Skills
Primarily indoors One location with some travel	Work Environment
Bachelor's degree	Minimum Education Level
$38,500 to $55,000 to $85,000	Salary Range
None available	Certification or Licensing
Little change or more slowly than average	Outlook

Overview

Metallurgical engineers develop new types of metal alloys and adapt existing materials to new uses. They manipulate the atomic and molecular structure of materials in controlled manufacturing environments, selecting materials with desirable mechanical, electrical, magnetic, chemical, and heat-transfer properties that meet specific performance requirements.

History

Metals weren't scientifically examined until the 19th century, but the roots of the science of metallurgy were developed more than 6,000 years before that. As far back as the Stone Age, when tools and weapons were being carved from rocks, people discovered that some rocks were actually nuggets of gold and could be used as a measure of value as well as for jewelry and ornaments.

By about 4300 BC, metals were being melted and molded into usable forms such as weapons. People then discovered that metals could be improved by mixing them with other components (such as blending copper and tin to form bronze). Such mixed metals are known as alloys. Metallurgical discoveries like this helped shape the flow of human civilization. After people discovered that copper could be melted to produce bronze, tougher weapons and tools were produced, thus changing aspects of warfare and power.

Rock deposits that contained metals became valuable, and people who had access to them wielded power. Such profitable mineral rock deposits came to be known as ores, and early alchemists developed methods for finding and preparing these ore deposits for metal extraction.

Iron has been an important metal extract since about 1200 BC, the beginning of the Iron Age. Alchemists refined smelting processes and began producing brass by combining copper and zinc, which was used to make coins in the Roman Empire. Throughout the next centuries, lead, silver, and gold (among other metals) continued to be mined, but the most significant developments in metallurgy focused on applications for iron. During the 18th and 19th centuries, metallurgists began to better understand the properties of metals. It was then that metallurgy as a science began.

Physical metallurgy, as a modern science, dates back to 1890, when a group of metallurgists began the study of alloys. Enormous advances were made in the 20th century, including the development of stainless steel, the discovery of a strong but lightweight aluminum, and the increased use of magnesium and its alloys. In recent years, metallurgical scientists have extended their research into nonmetallic materials, such as ceramics, glass, plastics, and semiconductors. This field has grown so broad that it is now often referred to as "materials science" to emphasize that it deals with both metallic and nonmetallic substances.

A relatively new area of metallurgy is powder metallurgy. Scientists have developed a process in which metals are turned into powders, compressed, and then heat-treated to produce a desired product. This method has resulted in the development of new alloys and composite materials.

Metallurgists are also concentrating on ways to reclaim and recycle solid wastes in order to conserve our natural resources and protect our environment. Many mineral-rich underground deposits have been depleted. Our bridges, buildings, and machines are made with metals that today have become more difficult to mine and more scarce than ever before. Metallurgical engineers are also focusing on issues concerning environmental protection (because extraction processes create pollution), recycling methods, and more efficient, automated processes of metal recovery, production, and reuse.

The Job

Metallurgical engineers are sometimes also referred to as *metallurgists*. There are basically three categories in which such engineers work. *Extractive metallurgists,* also known as *chemical metallurgists,* are concerned with the methods used to separate metals from ores, and the reclamation of materials from solid wastes for recycling. Among their responsibilities, they may supervise and control concentrating and refining processes in commercial mining operations. They may determine the methods used to concentrate the ore by separating minerals from dirt, rock, and other unwanted materials. Many of these separation methods are performed at a treatment plant or refinery. There the extractive metallurgist may supervise and control both the separation processes and final purification processes.

Extractive metallurgists also develop ways to improve the current methods of separating minerals. To do this, the extractive metallurgist processes small batches of ores in a laboratory and analyzes the efficiency of each operation and the feasibility of adapting the operations to commercial use. Extractive metallurgists also research ways to use new sources of metals, such as the reclamation of magnesium from seawater.

Extractive metallurgists often are involved in the design of treatment plants and refineries, and the equipment and processes used within them. They may determine the types of machines needed, supervise the installation of machinery, train refinery workers, and closely observe processing operations. They monitor operations and suggest new methods and modifications needed to improve efficiency.

Because minerals are becoming depleted in the environment, extractive metallurgical engineers are constantly searching for new ways to take metals from low-grade ores and to recycle metals that are considered scrap material. During the last 20 years, many of the refining processes have greatly improved, lessening environmental damage from waste materials.

Physical metallurgists, on the other hand, focus on the scientific study of the relationship between the structure and properties of metals and devise uses for metals. These engineers begin their job after metals have been extracted and refined. At that point, most such metals are not yet useful, so they must be improved by being blended with other metals and nonmetals to produce alloys.

Physical metallurgists may conduct X-ray and microscopic experiments on the metals to determine their physical structure and other characteristics, such as the amount of alloys and base metals present. These engineers also test the materials for impurities and defects and determine whether they can be used in thermal, electrical, or magnetic applications. The results of the studies and tests determine what the metal will be used for and how long it is expected to last.

Using the data gained during research, physical metallurgists also develop new applications for metals. They devise processes that transform the metals so they have desired characteristics such as hardness, corrosion resistance, malleability, and durability. These processes include hot working, cold working, foundry methods, powder metallurgy, nuclear metallurgy, and heat treatment. After the metals have been processed, they can be transformed into commercial products. *Metallographers* conduct the laboratory investigations on metal samples and prepare reports for physical metallurgists to evaluate.

Lastly, *process metallurgical engineers,* or *mechanical metallurgical engineers,* take metals and, by melting, casting, and mechanically processing them, produce forms that will be sold for a multitude of applications—such as automotive parts, satellite components, or coins. The field of process metallurgy is quite broad, involving such methods as welding, soldering, plating, rolling, and finishing metals to produce commercially standard products.

Requirements

High School

During high school, prospective metallurgical engineers should pursue a strong background in mathematics and physical sciences. At the very least, take chemistry and physics as well as algebra, geometry, and trigonometry. Computer science, analytical geometry, calculus, engineering science and design are also recommended.

Postsecondary Training

If your career goal is to become a metallurgical engineer, you will need a bachelor of science degree in materials or metallurgical engineering. Degrees are granted in many different specializations by more than 80 universities and colleges in the United States.

The first two years of college focus on subjects such as chemistry, physics, and mathematics, which are geared toward teaching analytical thinking. Students also take introductory engineering. By your sophomore year, you should have decided on a field of specialization because about one-third of your courses from then on will focus on metallurgy and related engineering areas.

There are a wide variety of programs available at colleges and universities, and it is helpful to explore as many of programs as possible that are accredited by the Accreditation Board for Engineering and Technology (ABET). Some programs prepare students for practical design and production work; others concentrate on theoretical science and mathematics.

More than 50 percent of metallurgical engineers begin their first job with a bachelor's degree. Many engineers continue on for a master's degree either immediately after graduation or after a few years of work experience. A master's degree generally takes two years of study. A doctoral degree requires at least four years of study and research beyond the bachelor's degree and is usually completed by engineers interested in research or teaching at the college level.

Other Requirements

If you are interested in metallurgical engineering, you should have a curiosity about how things work, an analytical mind, and mechanical ability. In general, metallurgical engineers are interested in nature and the physical sciences, and are creative and critical thinkers who enjoy problem solving. Engineers are patient, well organized, and attentive to detail because much of their work involves long-term projects and studies. They have good communications skills and are able to explain things easily to others. In addition, they can work comfortably both alone and with other people.

Exploring

Taking sculpture and welding classes is a good way of learning the properties of metals. Creating bronze sculptures, designing and making metal jewelry, and welding metals into structures provides hands-on experience. Interested high school students should read publications like the *JETS Report,* which is published by the Junior Engineering Technical Society (JETS). Science clubs such as JETS give students the opportunity to compete in academic events, take career exploration tests, and gain hands-on experience with metals and other materials.

Other excellent opportunities are found at summer camps and special academic programs. For example, Vanderbilt University has a summer program for high school students called Preparatory Academics for Vanderbilt Engineers (PAVE). Held in Nashville, Tennessee, PAVE offers activities in engineering, computer skills, problem solving, and technical writing. In college, students may join student chapters of associations such as the Society for Mining, Metallurgy, and Exploration (SME).

Employers

Opportunities for metallurgical engineers are found in a wide variety of settings, including metal-producing and processing companies, research institutes, and schools and universities. Engineers also work in aircraft manufacturing, machinery and electrical equipment manufacturing, the federal government, and for engineering consulting firms.

Starting Out

Most metallurgical engineers find their first job through their colleges' placement services. Technical recruiters visit universities and colleges annually to interview graduating students and possibly offer them jobs. Metallurgical engineers can also find work by directly applying to companies, through job listings at state and private employment services, or in classified advertisements in newspapers and trade publications.

Advancement

As in most occupations, the most experienced and educated workers stand the best chance for advancement. Metallurgical engineers with several years of technical experience are often eligible for supervisory positions; with further experience, engineers can apply for any number of managerial and administrative positions.

Engineers should keep current on technological advances in metallurgy throughout their careers. Many metallurgical engineers join professional associations, such as the Minerals, Metals, and Materials Society; the Society for Mining, Metallurgy, and Exploration; and the Metallurgical Society of the Canadian Institute of Mining, Metallurgy, and Petroleum. These associations hold annual conferences and meetings, in addition to other activities, which keep members up to date on recent developments and events within the industry. Special recognition—awards, scholarships, grants, and fellowships—are often given to those who demonstrate outstanding achievement in the field. For example, the Application to Practice Award is presented by the Minerals, Metals, and Materials Society to individuals who excel in translating their research work into practical manufacturing applications.

Earnings

According to the U.S. Bureau of Labor Statistics, the average annual salary for metallurgical engineers is about $55,000. Those with advanced degrees and experience can earn $85,000 or more per year, while recent graduates average about $38,500 annually. Benefits vary depending on the company, but usually include paid vacations and holidays, sick days, medical and dental insurance, profit sharing, and retirement plans. Some companies offer tuition assistance for continuing education and pay for membership and expenses for participation in professional associations.

Work Environment

Extractive metallurgical engineers usually work in ore treatment plants, refineries, smelter plants, or steel mills. They may also work at remote mining sites. Those working in physical metallurgy are usually located in labs or manufacturing plants, doing research and conducting studies on extracted metals. Process engineers work in a diverse range of environments, including welding shops, rolling mills, and industrial production plants for such products as automobiles and computer parts.

Those who choose research as their specialty will spend much time in labs and libraries. Those who work on school faculties will spend time in classrooms, but many are also employed by companies as working professional metallurgists.

Most metallurgical engineers work a 40-hour week. Metallurgical engineers who are employed in industrial refining may work on night shifts. Occasionally, evening or weekend work may be necessary to complete special projects or work on experiments.

Outlook

Employment for metallurgical engineers is expected to grow more slowly than average because many of the industries in which metallurgical engineers work are expected to have little if any employment growth through 2006. However, engineers should find sufficient job openings because of the low number of new graduates relative to other engineering disciplines.

Metallurgical engineers will increasingly work with companies that are developing new methods of processing low-grade ores—that is, those that have not yet been tapped because they are not as profitable as higher grades. As the world's ore deposits become further depleted, engineers will be needed to locate new sites and devise new alloy combinations. Also, metallurgical engineers will find jobs with companies that develop new methods of recycling scrap metals, and those that devise nonpolluting processing systems and cleanup methods for existing plants.

For More Information

For a list of accredited engineering programs at colleges and universities, contact:

Accreditation Board for Engineering and Technology, Inc.
111 Market Place, Suite 1050
Baltimore, MD 21202
Tel: 410-347-7700
Web: http://www.abet.ba.md.us

For an exceptionally useful Web site if you are considering engineering as a career, contact:

National Engineering Information Center
Web: http://www.asee.org/neic/

For information on careers, educational programs, and student chapters, contact:

Society for Mining, Metallurgy and Exploration
8307 Shaffer Parkway
Littleton, CO 80127
Tel: 800-763-3132
Email: smenet@aol.com
Web: http://www.smenet.org

For information on careers, educational programs, and scholarships, contact:

ASM International
9636 Kinsman Road
Materials Park, OH 44073-0002
Tel: 440-338-5151
Web: http://www.asm-intl.org

Junior Engineering Technical Society, Inc.
1420 King Street, Suite 405
Alexandria, VA 22314
Tel: 703-548-5387
Web: http://www.jets.org

The Minerals, Metals, & Materials Society (TMS)
184 Thorn Hill Road
Warrendale, PA 15086
Tel: 724-776-9000
Web: http://www.tms.org

Mining Engineers

Earth science Mathematics	School Subjects
Mechanical/manipulative Technical/scientific	Personal Skills
Indoors and outdoors Primarily multiple locations	Work Environment
Bachelor's degree	Minimum Education Level
$32,600 to $66,900 to $80,000+	Salary Range
Required by all states	Certification or Licensing
Decline	Outlook

Overview

Mining engineers deal with the exploration, location, and planning for removal of minerals and mineral deposits from the earth. These include metals (iron, copper), nonmetallic minerals (limestone, gypsum), and coal. Mining engineers conduct preliminary surveys of mineral deposits and examine them to ascertain whether they can be extracted efficiently and economically, using either underground or surface mining methods. They plan and design the development of mine shafts and tunnels, devise means of extracting minerals, and select the methods to be used in transporting the minerals to the surface. They supervise all mining operations and are responsible for mine safety. Mining engineers normally specialize in design, research and development, or production.

Mining equipment engineers may specialize in design, research, testing, or sales of equipment and services. Mines also require safety engineers.

History

The development of mining technology stretches back some 50,000 years to the period when people began digging pits and stripping surface cover in search of stone and flint for tools. Between 8000 and 3000 BC, the search for quality flint led people to sink shafts and drive galleries into limestone deposits.

By about 1300 BC, the Egyptians and other Near Eastern peoples were mining copper and gold by driving adits (near-horizontal entry tunnels) into hillsides, then sinking inclined shafts from which they drove extensive galleries. They supported the gallery roofs with pillars of uncut ore or wooden props.

Providing adequate ventilation posed a difficult problem in ancient underground mines. Because of the small dimensions of the passageways, air circulated poorly. All methods of ventilating the mines relied on the natural circulation of air by draft and convection. To assist this process, ancient engineers carefully calculated the number, location, and depth of the shafts. At the great Greek mining complex of Laurion, they sank shafts in pairs and drove parallel galleries from them with frequent crosscuts between galleries to assist air flow. Lighting a fire in one shaft caused a downdraft in the other.

Ancient Roman engineers made further advances in the mining techniques of the Greeks and Egyptians. They mined more ambitiously than the Greeks, sometimes exploiting as many as four levels by means of deep connecting shafts. Careful planning enabled them to drive complicated networks of exploratory galleries at various depths. Buckets of rock and ore could be hoisted up the main shaft by means of a windlass. Unlike the Greeks and Egyptians, the Romans often worked mines far below ground-water level. Engineers overcame the danger of flooding to some extent by developing effective, if expensive, drainage methods and machinery. Where terrain allowed, they devised an elaborate system of crosscuts to channel off the water. In addition, they adapted Archimedean screws—originally used for crop irrigation—to drain mine workings. A series of inclined screws, each emptying water into a tub emptied by a screw above it, could raise a considerable amount of water in a short time. It took only one man to rotate each screw, which made it perhaps the most efficient application of labor until engineers discovered the advantage of cutting halls large enough for an animal to rotate the screw. By the first century AD, the Romans had designed water wheels, which greatly increased the height to which water could be raised in mines.

Mining engineering advanced little from Roman times until the 11th century. From this period on, however, basic mining operations such as drainage, ventilation, and hoisting underwent increasing mechanization. In his book *De Re Metallica* (1556), the German scholar Georgius Agricola presented a detailed description of the devices and practices mining engineers had developed since ancient times. Drainage pumps in particular grew more and more sophisticated. One pump sucked water from mines by the movement of water-wheel-driven pistons.

As mines went deeper, technological problems required new engineering solutions. During the 18th century, engineers developed cheap, reliable steam-powered pumps to raise water in mines. Steam-powered windlasses also came into use. In the 1800s, engineers invented power drills for making shot holes for rock-breaking explosives. This greatly increased the capability to mine hard rock. In coal mines, revolving-wheel cutters—powered by steam, then by compressed air, then by electricity—relieved miners from the dangerous task of undercutting coal seams by hand. As late as the mid-19th century, ore was still being pushed or hauled through mines by people and animals. After 1900, however, electric locomotives, conveyor belts, and large-capacity rubber-tired vehicles came into wide use so that haulage could keep pace with mechanized ore breaking. The development of large, powerful machines also made possible the removal of vast amounts of material from open-pit mines.

The Job

Before the decision is made to mine a newly discovered mineral deposit, mining engineers must go through successive stages of information gathering, evaluation, and planning. As long as they judge the project to be economically viable, they proceed to the next stage. Review and planning for a major mining project may take a decade or longer and may cost many millions of dollars.

First mining engineers try to get a general idea of the deposit's potential. They accomplish this by reviewing geological data, product marketing information, and government requirements for permits, public hearings, and environmental protection. Based on this review, they prepare rough cost estimates and economic analyses. If it appears possible to mine the deposit at a competitive price with an acceptable return on investment, mining engineers undertake a more detailed review.

Meanwhile, geologists continue exploring the mineral deposit in order to ascertain its dimensions and character. Once the deposit has been reasonably well defined, mining engineers estimate the percentage of the deposit that can be profitably extracted. This estimate, which takes into account the ore's grade (value) and tonnage (volume and density), constitutes the minable ore reserve. It provides mining engineers with enough specific information to refine their economic appraisal and justify further analysis.

At this stage, engineers begin the process of selecting the most suitable mining method. They seek a method that will yield the largest profit consistent with safety and efficient ore extraction. In considering the adaptability of mining methods to the deposit, they rely heavily on rock-mechanics and geologic data. Measurements of the stresses, strains, and displacements in the rock surrounding the ore body help engineers predict roof-support requirements and settling of rock masses during excavation. Evaluation of the deposit's geologic features (such as the dimensions, inclination, strength, and physical character of the ore and overlying rock) enables engineers to place mine openings in stable rock, avoid underground water, and plan overall excavation procedures. If the evaluation calls for surface mining, engineers must decide where to dig the pits and where to put the rock and soil removed during mining.

Having estimated the ore reserve, chosen a mining method, and begun mine planning, engineers can determine daily (or yearly) mine output tonnage in light of product demand. They also select equipment and help plan and size the mine's plant, support, ore-processing, and shipping facilities.

For underground mining, mining engineers must determine the number and location of mine shafts, tunnels, and main extraction openings. They must also determine the size, number, kind, and layout of the various pieces of equipment. If the project continues to appear economically viable, construction begins.

As actual mine-making proceeds, mining engineers supervise operations. They train crews of workers and supervisors. The stress fields around the mine workings change as the mine expands. Engineers and engineering technicians must inspect the roof of underground cavities to ensure that it continues to have adequate support. Engineers must also continually monitor the quality of air in the mine to ensure proper ventilation. In addition, mining engineers inspect and repair mining equipment. Some mining engineers help plan ways of restoring or reclaiming the land around mine sites so that it can be used for other purposes.

Some mining engineers specialize in designing equipment used to excavate and operate mines. This equipment typically includes ventilation systems, earth- and rock-moving conveyors, and underground railroads and elevators. Engineers also design the equipment that chips and cuts rock and coal. Others select and determine the placement of explosives used to blast ore deposits.

Mining engineers also work for firms that sell mining supplies and equipment. Experienced mining engineers teach in colleges and universities and serve as independent consultants to industry and government.

Requirements

High School

To meet the standards set by most engineering colleges, high school students should take as much math and science as possible. Minimum course work includes elementary and intermediate algebra, plane geometry, trigonometry, chemistry, and physics. Courses in solid geometry, advanced algebra, and basic computer functions are highly recommended. In addition, many engineering colleges require three years of English (preferably emphasizing composition and public speaking) and social science (especially economics and history). Course work in foreign languages also is helpful, because many engineers work overseas.

Postsecondary Training

A bachelor's degree in engineering, preferably with a major in mining engineering, from an accredited engineering program is the minimum requirement for beginning mining engineering jobs. The organization that accredits engineering programs in the United States is the Accreditation Board for Engineering and Technology (ABET). ABET-accredited mining engineering programs assure students that their education will prepare them for professional practice or graduate study.

In a typical undergraduate engineering program, students spend the first two years studying basic sciences, such as mathematics, physics, and chemistry, as well as introductory engineering. Students must also study such subjects as economics, foreign languages, history, management, and writing. These courses equip students with skills they will need in their future work as engineers. The remaining years of college are devoted mostly to engineering courses, usually with a concentration in mining engineering. Engineering programs can last from four to six years. Those that require five to six years to complete may award a master's degree or provide a cooperative engineering education program. Cooperative programs allow students to combine classroom education and practical work experience with a participating mining company or engineering firm.

After completing their formal studies and landing a job, many mining engineers continue their education. They take courses, attend workshops, and read professional journals in order to keep up with developments in their field. Continuing education also enables them to acquire expertise in new technical areas. Some mining engineers pursue advanced degrees. A graduate degree is needed for most teaching and research positions and for many management positions. Some mining engineers pursue graduate study in engineering, business, or law.

Certification or Licensing

Regardless of their educational credentials, mining engineers normally must obtain professional certification in the states in which they work. Professional registration is mandatory for mining engineers whose work may affect life, health, or property or who offer their services to the public. Registration generally requires a degree from an ABET-accredited engineering program, four years of relevant work experience, and passing a state examination.

Other Requirements

Certain characteristics help qualify a person for a career in mining engineering. These include the judgment to adapt knowledge to practical purposes, the imagination and analytical skill to solve problems, and the capacity to predict the performance and cost of new processes or devices. Mining engineers must also be able to communicate effectively, work as part of a team, and supervise other workers.

Exploring

To learn about the profession of mining engineering, you may find it helpful to talk with science teachers and guidance counselors and with people employed in the minerals industry. You might also wish to read more about the industry and its engineers.

Companies and government agencies that employ graduates of mining engineering programs also hire undergraduates as part of a cooperative engineering education program. Students often enter such programs the summer preceding their junior year, after they have taken a certain number of engineering courses. They normally alternate terms of on-campus study and terms of work at the employer's facilities.

On the job, students assume the role of a junior mining engineer. They report to an experienced engineer, who acts as their supervisor and counselor. He or she assigns them work within their capabilities, evaluates their performance, and advises them as though they were permanent employees. Students have ample opportunity to interact with a diverse group of engineers and managers and to ask them about their work, their company, and mining engineering in general. Participation in the actual practice of the profession can help students assess their own aptitudes and interests and decide which courses will be most useful to them during the remainder of their engineering program.

Employers

In 1996, there were approximately 3,100 mining engineers employed in the United States. About two-thirds of them had jobs in the mining industry itself; the others worked for government agencies, manufacturing companies, or engineering consulting firms.

Starting Out

Beginning mining engineers generally perform routine tasks under the supervision of experienced engineers. Some mining companies provide starting engineers with in-house training. As engineers gain knowledge and experience, they receive increasingly difficult assignments along with greater independence to develop designs, solve problems, and make decisions.

Advancement

Mining engineers may become directors of specific mining projects. Some head research projects. Mining engineers may go on to work as technical specialists or to supervise a team of engineers and technicians. Some eventually manage their mining company's engineering department or enter other managerial, management support, or sales positions.

Earnings

During the 1990s, salaries for mining engineers with a bachelor's degree averaged $32,638 a year. The median average income for mining engineers is $66,885 a year. Mining engineers with considerable experience who enter management positions may command salaries of $80,000 or more. Engineers who work for the federal government in its mining operations tend to earn slightly less than their counterparts in private industry.

Work Environment

Engineers in the mining industry generally work where the mineral deposits are situated, often near small, rural communities. But those who specialize in research, management, consulting, or sales may work in metropolitan areas.

For those who work at the mine sites, conditions vary depending on the mine's location and structure and on what the engineer does. Conditions in the underground environment differ from those in surface mining. Natural light and fresh air are absent; temperatures may be uncomfortably hot or cold. Some mines have large amounts of water seeping into the openings. Potential hazards include caving ground, rockfalls, explosions from accumulation of gas or misuse of explosives, and poisonous gases. Most mines, however, are relatively safe and comfortable, owing to artificial light and ventilation, protective clothing, and water-pumping and ground-support systems.

Many mining engineers work a standard 40-hour week. In order to meet project deadlines, however, they may have to work longer hours under considerable stress.

Outlook

The demand for mining engineers is expected to decline through 2006 because of predicted low growth in the demand for coal, metals, and other minerals as long as energy sources other than coal—petroleum, natural gas, and nuclear energy—are readily available at more reasonable prices. The employment rate for mining engineers in the United States also depends on the price of coal and metals from other countries. A certain number of mining engineers, however, will always be needed. As mineral deposits are depleted, engineers will have to devise ways of mining less accessible low-grade ores to meet the demand for new alloys and new uses for minerals and metals. More technologically advanced mining systems and more stringently enforced health and safety regulations may also have a positive effect on the employment of mining engineers. Job openings also result from the need to replace those who transfer to specialized jobs within the field or to other occupations.

For More Information

For information on careers, schools, college student membership, scholarships and grants, and other resources, contact:

The Minerals, Metals, & Materials Society (TMS)
184 Thorn Hill Road
Warrendale, PA 15086
TMS Education Department
Tel: 724-776-9000
Web: http://www.tms.org

This organization is a merger of the National Coal Association and the American Mining Congress. It is an information center for the mining industry.

National Mining Association
1130 17th Street, NW
Washington, DC 20036
Tel: 202-463-2625
Web: http://www.nma.org

For career information, contact:

Society for Mining, Metallurgy and Exploration
Career Information Department
8307 Shaffer Parkway
Littleton, CO 80127
Tel: 303-973-9550
Web: http://www.smenet.org

Nuclear Engineers

	School Subjects
Mathematics Physics	

	Personal Skills
Communication/ideas Technical/scientific	

	Work Environment
Primarily indoors Primarily one location	

	Minimum Education Level
Bachelor's degree	

	Salary Range
$33,600 to $65,000 to $100,000	

	Certification or Licensing
Required for certain positions	

	Outlook
Little change or more slowly than the average	

Overview

Nuclear engineers are concerned with accessing, using, and controlling the energy released when the nucleus of an atom is split. The process of splitting atoms, called fission, produces a nuclear reaction, which creates radiation in addition to nuclear energy. Nuclear energy and radiation has many uses. Some engineers design, develop, and operate nuclear power plants, which are used to generate electricity and power navy ships. Others specialize in developing nuclear weapons, medical uses for radioactive materials, and disposal facilities for radioactive waste.

History

Nuclear engineering as a formal science is quite young. However, part of its theoretical foundation rests with the ancient Greeks. In the fifth century BC, Greek philosophers postulated that the building blocks of all matter were indestructible elements, which they named *atomos,* meaning "indivisible."

This atomic theory was accepted for centuries, until the British chemist and physicist John Dalton (1766-1844) revised it in the early 1800s. In the following century, scientific and mathematical experimentation led to the formation of modern atomic and nuclear theory.

Today, it is known that the atom is far from indivisible and that its dense center, the nucleus, can be split to create tremendous energy. The first occurrence of this splitting process was inadvertently induced in 1938 by two German chemists. Further studies confirmed this process and established that the fragments resulting from the fission in turn strike the nuclei of other atoms, resulting in a chain reaction that produces constant energy.

The discipline of modern nuclear engineering is traced to 1942, when physicist Enrico Fermi (1901-1954) and his colleagues produced the first self-sustained nuclear chain reaction in the first nuclear reactor ever built. In 1950, North Carolina State College offered the first accredited nuclear engineering program. By 1965, nuclear engineering programs had become widely available at universities and colleges throughout the country and worldwide. These programs provided engineers with a background in reactor physics and control, heat transfer, radiation effects, and shielding.

Current applications in the discipline of nuclear engineering include the use of reactors to propel naval vessels and the production of radioisotopes for medical purposes. Most of the growth in the nuclear industry, however, has focused on the production of electric energy.

Despite the controversy over the risks involved with atomic power, in 1990 France derived 75 percent of its electricity from nuclear power; Germany, 60 percent; and the United States, 21 percent. Medicine, manufacturing, and agriculture have also benefited from nuclear research. Such use requires the continued development of nuclear waste management. Low-level wastes, which result from power plants as well as hospitals and research facilities, must be reduced in volume, packed in leak-proof containers, and buried, and waste sites must be continually monitored.

The Job

Nuclear engineers are involved in various aspects of the generation, use, and maintenance of nuclear energy and the safe disposal of its waste. Nuclear engineers work on research and development, design, fuel management, safety analysis, operation and testing, sales, and education. Their contributions affect consumer and industrial power supplies, medical technology, the food industry, and other industries.

Nuclear engineering is dominated by the power industry. Some engineers work for companies that manufacture reactors. They research, develop, design, manufacture, and install parts used in these facilities, such as core supports, reflectors, thermal shields, biological shields, instrumentation, and safety and control systems.

Those who are responsible for the maintenance of power plants must monitor operations efficiently and guarantee that facilities meet safety standards. Nuclear energy activities in the United States are closely supervised and regulated by government and independent agencies, especially the Nuclear Regulatory Commission (NRC). The NRC oversees the use of nuclear materials by electric utility companies throughout the United States. NRC employees are responsible for ensuring the safety of nongovernment nuclear materials and facilities and for making sure that related operations do not adversely affect public health or the environment. Nuclear engineers who work for regulatory agencies are responsible for setting the standards that all organizations involved with nuclear energy must follow. They issue licenses, establish rules, implement safety research, perform risk analyses, conduct on-site inspections, and pursue research. The NRC is one of the main regulatory agencies employing nuclear engineers.

Many nuclear engineers work directly with public electric utility companies. Tasks are diverse, and teams of engineers are responsible for supervising construction and operation, analyzing safety, managing fuel, assessing environmental impact, training personnel, managing the plant, storing spent fuel, managing waste, and analyzing economic factors.

Some engineers working for nuclear power plants focus on the quality of the water supply. Their plants extract salt from water, and engineers develop new methods and designs for such desalinization systems.

The food supply also benefits from the work of nuclear engineers. Nuclear energy is used for pasteurization and sterilization, insect pest control, and fertilizer production. Furthermore, nuclear engineers conduct genetic research on improving various food strains and their resistance to harmful elements.

Nuclear engineers in the medical field design and construct equipment for diagnosing and treating illnesses and disease. They perform research on radioisotopes, which are produced by nuclear reactions. Radioisotopes are used in heart pacemakers, X-ray equipment, and for sterilizing medical instruments. Fifty percent of all U.S. hospital patients benefit from a procedure or device that uses radioisotopes.

Numerous other jobs are performed by nuclear engineers. *Nuclear health physicists, nuclear criticality safety engineers,* and *radiation protection technicians* conduct research and training programs designed to protect plant and laboratory employees against radiation hazards. *Nuclear fuels research engineers* and *nuclear fuels reclamation engineers* work with reprocessing systems for atomic fuels. *Accelerator operators* coordinate the operation of equipment used in experiments on subatomic particles, and *scanners* work with photographs, produced by particle detectors, of atomic collisions.

Requirements

High School

If you are interested in becoming a professional engineer, you must begin preparing yourself in high school. You should take honors-level courses in mathematics and the sciences. Specifically, you should complete courses in algebra, geometry, trigonometry, and calculus; chemistry, physics, and biology; English, social studies, foreign language (many published technical papers that are required reading in later years are written in German or French) and humanities; and computer science.

Postsecondary Training

Professional engineers must have at least a bachelor's degree. You should attend a four-year college or university that is approved by the Accreditation Board for Engineering and Technology. In a nuclear engineering program, you will focus on subjects similar to those studied in high school but at a more advanced level. Courses also include engineering sciences and atomic and nuclear physics.

These subjects will prepare you for analyzing and designing nuclear systems and understanding how they operate. You will learn and comprehend what is involved in the interaction between radiation and matter; radiation measurements; the production and use of radioisotopes; reactor physics and engineering; and fusion reactions. The subject of safety will be emphasized, particularly with regards to handling radiation sources and implementing nuclear systems.

You must have a master's or doctoral degree for most jobs in research, higher education, and for supervisory and administrative positions. It is recommended that you obtain a graduate degree in nuclear engineering because this level of education will help you obtain the skills required for advanced specialization in the field. Many institutions that offer advanced degrees have nuclear reactors and well-equipped laboratories for teaching and research. You can obtain information about these schools by contacting the U.S. Department of Energy.

Certification or Licensing

A Professional Engineer (P.E.) license is usually required before obtaining employment on public projects (i.e., work that may affect life, health, or property). Although registration guidelines differ for each state, most states require a degree from an accredited engineering program, four years of work experience in the field, and a minimum grade on a state exam.

Other Requirements

Nuclear engineering occupations will encounter two unique concerns. First, exposure to high levels of radiation may be hazardous; thus, engineers must always follow safety measures. Those working near radioactive materials must adhere to strict precautions outlined by regulatory standards. In addition, female engineers of childbearing age may not be allowed to work in certain areas or perform certain duties because of the potential harm to the human fetus from radiation.

Finally, nuclear engineers must be prepared for a lifetime of continuing education. Because nuclear engineering is founded in the fundamental theories of physics and the notions of atomic and nuclear theory are difficult to conceptualize except mathematically, an aptitude for physics, mathematics, and chemistry is indispensable.

Exploring

Each year, the U.S. Department of Energy (DOE) sponsors a program for high school science honor students. You are eligible for this program if you are 16 to 18 years of age and have finished the 11th grade. This program has focused on the areas of high-energy particle physics, computational sciences using the Cray II supercomputer, and materials science involving superconductivity. The DOE pays for round-trip airfare, lodging, meals, recreational activities, and special awards. If you are interested in this program, contact your school's student honors program.

If you are in the top 10 percent of your class and between 11 and 17 years of age, you are eligible for Clemson University's summer science and engineering honors program. In this program, you will conduct lab experiments and participate in field trips related to subjects such as physics, biology, and creativity in engineering design. For more information, write to: Clemson University, Suite 532, Clemson House, Clemson, SC 29634.

If you are interested in becoming an engineer, you can join science clubs such as the Junior Engineering Technical Society (JETS), which has a chapter in almost every state. Science clubs provide the opportunity to work with others, design engineering projects, and participate in career exploration. The *JETS Report* will introduce you to the organization and includes articles about club activities and student interests. If you are a more advanced student, you may want to read *ANS News,* published by the American Nuclear Society. This publication reports monthly on the society's members and activities.

Employers

Nuclear engineers work in a variety of settings. In 1996, the following three sectors each represented about 20 percent of nuclear engineers: utilities, the federal government, and engineering consulting firms. Another 12 percent were in research and testing services. More than half of all federally employed nuclear engineers were civilian employees of the navy, and most of the rest worked for the Nuclear Regulatory Commission, the Department of Energy, or the Tennessee Valley Authority. Most nonfederally employed nuclear engineers worked for public utilities or engineering consulting companies. Some worked for defense manufacturers or manufacturers of nuclear power equipment.

Starting Out

Most students begin their job search while still in college, collecting advice from job counselors and their school's placement center. Also, the DOE has training programs that help applicants qualify for positions in nuclear engineering. For information, contact the agency's Office of Personnel

Management. The Society of Women Engineers also administers several certificate and scholarship programs and advises students regarding job placements.

As with other engineering disciplines, a hierarchy of workers exists, with the chief engineer having overall authority over managers and project engineers. This is true whether you are working in research, design, production, sales, or teaching. After gaining a certain amount of experience, engineers may apply for positions in supervision and management.

Advancement

Because the nuclear engineering field is so young, the time is ripe for technological developments, and engineers must therefore keep abreast of new research and technology throughout their careers. Advancement for engineers is contingent upon continuing education, research activity, and on-the-job expertise.

Advancement may also bring recognition in the form of grants, scholarships, fellowships, and awards. For example, the American Nuclear Society has established a Young Members Engineering Achievement Award to recognize outstanding work performed by members. To be eligible for this award, you must be younger than 40 years and demonstrate effective application of engineering knowledge that results in a concept, design, analysis method, or product used in nuclear power research and development or in a manufacturing application.

Earnings

Entry-level salaries for engineers are usually higher than those earned in any other occupations. Nuclear engineers earn salaries approximately midway between those earned by the highest paid engineers (petroleum) and the lowest paid engineers (civil). In 1994, nuclear engineers with a bachelor's degree earned $33,600 in their first year on the job.

If you have a master's or a doctoral degree, you may earn a starting salary that is 10 to 15 percent higher than if you have only a bachelor's degree. After more than 5 years in the industry, nuclear engineers usually see a rise in their salary by an average of $5,000 to $8,000; after 10 years, salaries rise by an average of $15,000. Nuclear engineers with considerable training and experience may earn more than $100,000 a year.

Work Environment

In general, nuclear engineering is a technically demanding and politically volatile field. Those who work daily at power plants perhaps incur the most stress because they are responsible for preventing large-scale accidents involving radiation. Those who work directly with nuclear energy face risks associated with radiation contamination. Engineers handling the disposal of hazardous material also work under stressful conditions because they must take tremendous care to ensure the public's health and safety.

Research, teaching, and design occupations allow engineers to work in laboratories, classrooms, and industrial manufacturing facilities. Many engineers who are not directly involved with the physical maintenance of nuclear facilities spend most of their working hours, an average of 46 hours per week, conducting research. Most work at desks and must have the ability to concentrate on very detailed data for long periods of time, drawing up plans and constructing models of nuclear applications.

Outlook

In 1996, nuclear engineers held about 14,000 jobs. According to the U.S. Department of Labor, the number of new jobs is expected to grow more slowly than the average for all occupations through 2006. Most openings will arise as nuclear engineers transfer to other occupations or leave the labor force. However, good opportunities for nuclear engineers should still exist because the small number of nuclear engineering graduates is likely to be in balance with the number of job openings.

Because of public concern about the cost and safety of nuclear power, there are only a few nuclear power plants under construction in the United States, and it is possible some older plants will shut down. Nevertheless, nuclear engineers will be needed to operate existing plants. They will also continue to be needed to work in defense-related areas, to develop nuclear medical technology, and to improve and enforce waste management and safety standards.

For More Information

This nonprofit, international, scientific and educational organization is made up of 13,000 members who promote the advancement of engineering and science relating to the atomic nucleus, and of allied sciences and arts. The society provides information, publications, scholarships, and seminars, and cooperates in educational efforts.

American Nuclear Society
555 North Kensington Avenue
LaGrange Park, IL 60526
Tel: 708-352-6611
Web: http://www.ans.org

These organizations advocate peaceful use of nuclear technologies.

Nuclear Energy Institute
1776 I Street, NW, Suite 400
Washington, DC 20006
Tel: 202-739-8000
Web: http://www.nei.org/introstore.html

American Society for Nondestructive Testing
1711 Arlingate Lane
Columbus, OH 43228
Tel: 614-274-6003
Web: http://www.asnt.org

For career guidance and scholarship information, contact:

The Society of Women Engineers
345 East 47th Street, Room 305
New York, NY 10017
Tel: 212-509-9577
Web: http://www.swe.org

Optical Engineers

Overview

Optical engineers apply the concepts of optics to research, design, and develop applications in a broad range of areas. Optics, which involves the properties of light and how it interacts with matter, is a branch of physics and engineering. Optical engineers utilize their knowledge of methods by which light is produced, transmitted, detected, and measured to determine ways it can be used and to build devices using optical technology.

History

The study of the properties of light began during the 1600s when Galileo built telescopes to observe the planets and stars. Scientists, such as Sir Isaac Newton, conducted experiments and studies that contributed to the understanding of light and how it operates. Among Newton's many experiments is work done with prisms to break sunlight into a spectrum of colors. Important studies also were done by Christiaan Huygens, a Dutch physicist, who came up with a theory that concerned the wave properties of light.

During the 1800s, other physicists and scientists performed research that confirmed Huygens' theory and advanced the study of light even further. By the mid-1800s, scientists were able to measure the speed of light and had developed means to show how color bands of the light spectrum were created by atoms of chemical elements. In 1864, a British physicist, James C. Maxwell, proposed the electromagnetic theory of light.

Scientists during the 1800s and 1900s made several other important discoveries that advanced the understanding of light. Two of the most important discoveries are the development of lasers and fiber optics. The first laser was built by an American physicist, Theodore H. Maiman, in 1960. In 1966, it was discovered that light could travel through glass fibers and this led to the development of fiber optic technology.

Optics, the branch of science that studies the manipulation of light, is a growing field, and engineers today work in applications that include image processing, information processing, wireless communications, audio/CD players, high-definition televisions, laser printers, astronomical observation, atomic research, robotics, military surveillance, water-quality monitoring, undersea monitoring, and medical and scientific procedures and instruments.

The Job

Optical engineers may work in any of the many subfields or related branches of optics. Three of the largest areas are physical optics, which is concerned with the wave properties of light; quantum optics, which studies the photons, or individual particles of light; and geometrical optics, which involves optical instruments used to detect and measure light. Other subfields of optics include integrated optics, nonlinear optics, electron optics, magneto-optics, and space optics.

Optical engineers combine their knowledge of optics with other engineering concepts, such as mechanical engineering, electrical engineering, and computer engineering, to determine applications for optical technology and to build devices using this technology. Optical engineers may design optical systems for precision optical systems such as cameras, telescopes, or lens systems. They determine the required specifications and make adjustments to calibrate and fine tune optical devices. They also design and develop circuitry and components for devices that use optical technology. Some optical engineers design and fabricate inspection instruments to test and measure the performance of optical systems. In designing this equipment, they test that all parts perform as required, diagnose any malfunctioning parts, and correct any defects found. Together with electrical and mechanical engineers, they work on the overall design of systems using optical components.

In creating a new product using optical technology, optical engineers go through a multistep engineering process. First, they study the application or problem to understand it thoroughly. Then they brainstorm or use their imaginations to come up with possible solutions to the problem. Once they come up with a creative concept, they transform it into a design or several designs. They work out all of the details of the design and then create a computer-generated model or test unit. This model or unit is tested and any required revisions to the design are made. A revised unit or model is built and tested. This process continues until the design proves satisfactory. The design is then sent to production, and the product is manufactured. The process is completed with marketing of the product.

For some products, an engineer may perform all of the steps except for the marketing step. Other products require a team of engineers and may include other professionals such as industrial designers, technologists, and technicians.

Some optical engineers specialize in lasers and fiber optics. These engineers, also known as *fiber optics engineers* and *laser and fiber optics engineers,* design, develop, modify, and build equipment and components that utilize laser and fiber optic technology. Fiber optics are thin, hairlike strands of plastic-coated glass fibers that transmit light and images. Lasers may be used to generate the light in these fibers. Lasers, which are devices that produce extremely thin, powerful beams of light, have special properties that allow them to be used in many applications. Lasers can cut through material as hard as a diamond, can travel over long distances without any loss of power, and can detect extremely small movements. Lasers also can be used to record, store, and transmit information.

These engineers may be involved in testing laser systems or developing applications for lasers in areas such as telecommunications, medicine, defense, manufacturing, and construction. For example, lasers can be used in surgical procedures and medical diagnostic equipment. Lasers also are used in manufacturing industries to align, mark, and cut through both metals and plastics. Lasers are used in military applications such as navigational systems and ballistic and weapon systems. Other areas where optical engineers use lasers include robotics, holograms, scanning, compact discs, and printing.

Fiber optics engineers may specialize and work within a specific area of fiber optic technology. They may work with fiber optic imaging, which involves using fiber optics to transmit light or images. These engineers also work with fiber optics to rotate, enlarge, shrink, and enhance images. A second area of specialization is work with sensors. These engineers work with devices that measure temperature, pressure, force, and other physical features. Communications is a third type of specialization involving fiber optics. Fiber optic networks allow voice, data, sound, and images to be transmitted over cables. This area involves telephone systems, computer networks, and undersea fiber optic communications systems.

Optical engineers use many different types of equipment to perform their work. Among them are spectrometers, spectrum analyzers, digital energy meters and calorimeters, laser power meters, leak detectors, and wattmeters.

Requirements

High School

High school students should take physical science, physics, chemistry, geometry, algebra, trigonometry, calculus, social studies, English, composition, and computer science classes. Courses in computer-aided design are also helpful. Honors classes in science and mathematics are recommended. If you are planning to pursue an advanced degree beyond a bachelor's you will also need to take or be proficient in a foreign language.

Postsecondary Training

A bachelor of science degree in engineering is required to become an optical engineer. Most engineering programs take four or five years to complete. Many students also receive advanced degrees, such as a master of science degree or a doctoral degree, as these degrees are required for the best job opportunities.

There are about 120 colleges and universities in the United States and approximately five in Canada that offer classes in optics. Only a very small number of schools, though, offer programs that grant degrees in optical engineering. Most colleges offer degrees in a related field, such as electrical engineering or physics, with a specialization or option in optics.

Colleges may offer optical engineering classes through various departments, such as physics, electrical and computer engineering, electronic and electrical engineering, eptoelectronics and photonics imaging, optical engineering, or optical science. Each college's program is unique. Some schools emphasize the engineering aspects of optics, whereas others focus on optical science, or the research aspects of optics. Optical science varies from optical engineering in that it is more concerned with studying and understanding the characteristics and properties of light and its interaction with matter than in developing applications that utilize optical technology.

Classes vary based on the type of program, but generally include intensive laboratory experience and courses in mathematics, physics, chemistry, electronics, and geometric and wave optics. Advanced courses may include electro-optics, lasers, optical systems design, infrared systems design, quantum mechanics, polarization, fiber optics communication, and optical tests and measurement.

Some colleges require internships or cooperative work programs during which students work at a related job for one to three semesters. Alternating study with work experience is an excellent way to gain on-the-job experience before graduation and can lead to employment opportunities upon graduation.

A high number of students receive master of science degrees, which generally take two years of additional study beyond a bachelor's degree. Those who plan to work in research generally earn doctoral degrees, which take four years of additional study beyond a bachelor's degree.

Because the types of programs vary, you should thoroughly research and investigate as many colleges as possible. SPIE, the International Society for Optical Engineering, provides a detailed annual directory that lists colleges and universities offering optics courses and describes programs and requirements in depth.

Other Requirements

To become an optical engineer you need to have a strong foundation in mathematics and physics, as well as an inquisitive and analytical mind. You should be good at problem solving, enjoy challenges, and be methodical, precise, and attentive to details. You should be able to work well both individually and with others.

Exploring

Students interested in optics can join science and engineering clubs that provide opportunities for experimentation, problem-solving, and teambuilding activities. These clubs provide good grounding in science and math principles and the skills students will need as engineers. You may be able to arrange independent study projects with a science teacher. Another possibility is to perform simple experiments that examine the properties of light. Books on optics often provide instructions for experiments that may be done with a minimum of equipment. Books may be found at a public library or through the help of a science teacher.

A career video is available through SPIE. The video shows how optics are applied in various fields and explains the necessary steps to enter these fields. To receive this video, write to SPIE. A small fee is charged for duplication and shipping costs.

College students may wish to consider joining a student chapter of a professional association such as SPIE or the Optical Society of America. Participation in association events provides an excellent means to meet with professionals working in the area of optical engineering and to learn more about the field. In addition, membership may include a subscription to trade magazines that include interesting and informative articles on optics. Although these associations do charge membership fees, they are relatively inexpensive for college students.

Employers

Optical engineers work for companies that produce robotics. They also work in laboratories, hospitals, and universities, as well as medicine, telecommunications, and construction. Companies that employ optical engineers can be found in all areas of the country, although some areas have a higher concentration of such companies than others. Areas with large numbers of companies that employ optical engineers can be found along the Atlantic coast, from Boston to Washington, DC, and in large metropolitan areas around cities such as San Jose, Los Angeles, Dallas, Houston, and Orlando.

Starting Out

Some students work part time or during the summer during their college years as laser technicians, optics technicians, or related types of technicians. This employment may help them to receive an offer for full-time employment as an optical engineer once they have completed their education. It also provides a valuable learning experience and may help students decide the area of optics in which they would like to work.

Students in an undergraduate or graduate program may learn about job openings through internships or cooperative programs in which they have participated. Job leads also may be obtained through their colleges' job placement offices. Professional associations also may provide information on companies that are seeking optical engineers. In many cases, students will need to research companies that hire optical engineers and then apply directly to them.

Advancement

Optical engineers with a bachelor of science degree often start out as assistants to experienced engineers. As they gain experience, they are given more responsibility and independence and move into higher-ranked positions. Engineers who show leadership ability, good communication skills, and management ability may advance to project engineers, project managers, team leaders, or other management positions.

Engineers with bachelor's degrees often return to school to obtain advanced degrees, either an M.S. or Ph.D. With advanced training and experience, they can move into more specialized areas of engineering. Some engineers move into areas of research and become principal engineers or research directors. Engineers may also become college professors or high school teachers.

Some engineers move into sales and marketing. Selling optical devices requires a depth of technical knowledge and the ability to explain the features and benefits of a product. Many engineers, after having spent years designing products, are well equipped for this type of work.

Other optical engineers go into business for themselves, either becoming consulting engineers or starting their own design or manufacturing firms.

Earnings

Salaries for optical engineers are similar to those of electrical and electronics engineers. Entry-level engineers generally earn an average salary of $39,000. Engineers with 10 or more years of experience earn about $55,000 to $89,000, or more, a year.

Companies offer a variety of benefits including the following: medical, dental, and vision insurance; paid holidays, paid vacations, sick leave, and personal days; life and disability insurance; pension plans; profit sharing; 401(k) plans; tuition assistance programs; and release time for additional education. Some companies also pay for fees and expenses to participate in professional associations, including out-of-town travel to national conventions, annual meetings, and trade shows.

Work Environment

Optical engineers generally work in comfortable surroundings. They usually work in an office or laboratory. Most facilities are equipped with modern equipment and computer workstations. Most engineers work five-day, 40-hour weeks, although overtime is not unusual, particularly when working on a special project. Some companies offer flexible work policies in which engineers can schedule their own hours within certain time periods. Most engineers work with other engineers, technicians, and production personnel.

Outlook

Opportunities for optical engineers are very good and should remain high in the coming decade. At present, there are more openings for qualified engineers than there are available engineers to fill these positions.

Applications that utilize optics technology are growing steadily and should provide opportunities in many different industries. The use of fiber optics in telecommunications is expanding, providing opportunities for engineers in the cable, broadcasting, computer, and telephone industries. New applications are being developed in many other areas, such as the medical and defense fields. The increasing use of automated equipment in manufacturing is also providing opportunities for optical engineers, particularly in applications involving robotics technology.

For More Information

For information on careers and student membership, contact:

Lasers and Electro-Optics Society
c/o The Institute of Electrical and Electronics Engineers
PO Box 1331
445 Hoes Lane
Piscataway, NJ 08855-1331
Tel: 732-562-3892

To receive a directory of colleges or a career video, or for information on scholarships and student membership, contact:

SPIE—The International Society for Optical Engineering
PO Box 10
1000 20th Street
Bellingham, WA 98227-0010
Tel: 360-676-3290
Web: http://www.spie.org

For information on student membership, contact:

Optical Society of America
2010 Massachusetts Avenue, NW
Washington, DC 20036
Tel: 202-223-8130
Web: http://www.osa.org

Packaging Engineers

Mathematics Physics	School Subjects
Mechanical/manipulative Technical/scientific	Personal Skills
Primarily indoors Primarily one location	Work Environment
Bachelor's degree	Minimum Education Level
$35,000 to $45,000 to $60,000+	Salary Range
Required by certain states	Certification or Licensing
Faster than the average	Outlook

Overview

The *packaging engineer* designs, develops, and specifies containers for all types of goods, such as food, clothing, medicine, housewares, toys, electronics, appliances, and computers. In creating these containers, some of the packaging engineer's activities include product and cost analysis, management of packaging personnel, development and operation of packaging filling lines, and negotiations with customers or sales representatives.

Packaging engineers may also select, design, and develop the machinery used for packaging operations. They may either modify existing machinery or design new machinery to be used for packaging operations.

History

Certain packages, particularly glass containers, have been used for over 3,000 years; the metal can was developed to provide food for Napoleon's army. However, the growth of the packaging industry developed during the Industrial Revolution, when shipping and storage containers were needed for the increased numbers of goods produced. As the shipping distance from producer to consumer grew, more care had to be given to packaging so goods would not be damaged in transit. Also, storage and safety factors became important with the longer shelf life required for goods produced.

Modern packaging methods have developed since the 1920s with the introduction of cellophane wrappings. Since World War II, early packaging materials such as cloth and wood have been largely replaced by less expensive and more durable materials such as steel, aluminum, and plastics such as polystyrene. Modern production methods have also allowed for the low-cost, mass production of traditional materials such as glass and paperboard. Both government agencies and manufacturers and designers are constantly trying to improve packaging so that it is more convenient, safe, and informative.

Today, packaging engineers must also consider environmental factors when designing packaging because the disposal of used packages has presented a serious problem for many communities. The United States uses more than 500 billion packages yearly; 50 percent of these are used for food and beverages, and another 40 percent for other consumer goods. To help solve this problem, the packaging engineer attempts to come up with solutions such as the use of recyclable, biodegradable, or less bulky packaging.

The Job

Packaging engineers plan, design, develop, and produce containers for all types of products. When developing a package, they must first determine the purpose of the packaging and the needs of the end-user and their clients. Packaging for a product may be needed for a variety of reasons: for shipping, storage, display, or protection. A package for display must be attractive as well as durable and easy to store; labeling and perishability are important considerations, especially for food, medicine, and cosmetics. If the packaging purpose is for storage and shipping, then ease of handling and durability have to be considered. Safety factors may be involved if the materials to be packaged are hazardous, such as toxic chemicals or explosives. Finally, the costs of producing and implementing the packaging have to be considered, as well as the packaging material's impact on the environment.

After determining the purpose of the packaging, the engineers study the physical properties and handling requirements of the product in order to develop the best kind of packaging. They study drawings and descriptions of the product or the actual product itself to learn about its size, shape, weight, and color, the materials used, and the way it functions. They decide what kind of packaging material to use and with the help of designers and production workers, they make sketches, draw up plans, and make samples of the package. These samples, along with lists of materials and cost estimates, are submitted to management or directly to the customer. Computer design programs and other related software may be used in the packaging design and manufacturing process.

When finalizing plans for packaging a product, packaging engineers contribute additional expertise in other areas. They are concerned with efficient use of raw materials and production facilities as well as conservation of energy and reduction of costs. For instance, they may use materials that can be recycled, or they may try to cut down on weight and size. They must keep up with the latest developments in packaging methods and materials and often recommend innovative ways to package products. Once all the details for packaging are worked out, packaging engineers may be involved in supervising the filling and packing operations, operating production lines, or drawing up contracts with customers or sales representatives. They should be knowledgeable about production and manufacturing processes, as well as sales and customer service.

After a packaging sample is approved, packaging engineers may supervise the testing of the package. This may involve simulation of all the various conditions a packaged good may be subjected to, such as temperature, handling, and shipping.

This can be a complex operation involving several steps. For instance, perishable items such as food and beverages have to be packaged to avoid spoilage. Electronic components have to be packaged to prevent damage to parts. Whether the items to be packaged are food, chemicals, medicine, electronics, or factory parts, considerable knowledge of the properties of these products is often necessary to make suitable packaging.

Design and marketing factors also need to be considered when creating the actual package that will be seen by the consumer. Packaging engineers work with *graphic designers* and *packaging designers* to design effective packaging that will appeal to consumers. For this task, knowledge of marketing, design, and advertising are essential. Packaging designers consider color, shape, and convenience as well as labeling and other informative features when designing packages for display. Very often, the consumer is able to evaluate a product only from its package.

The many different kinds of packages require different kinds of machinery and skills. For example, the beverage industry produces billions of cans, bottles, and cardboard containers. Often packaging engineers are involved in selecting and designing packaging machinery along with other engineers and production personnel. Packaging can be manufactured either at the same facility where the goods are produced or at facilities that specialize in producing packaging materials.

The packaging engineer must also consider safety, health, and legal factors when designing and producing packaging. Various guidelines apply to the packaging process of certain products and the packaging engineer must be aware of these regulations. Labeling and packaging of products are regulated by various federal agencies such as the Federal Trade Commission and the Food and Drug Administration. For example, the Consumer Product Safety Commission requires that safe packaging materials be used for food and cosmetics.

Requirements

High School

During high school, students planning to enter a college engineering or packaging program should take college algebra, trigonometry, physics, chemistry, computer science, mechanical drawing, economics, and accounting classes. Speech, writing, art, computer-aided design, and graphic arts classes are also recommended.

Postsecondary Training

Several colleges and universities offer a major in packaging engineering. These programs may be offered through an engineering school or a school of packaging within a university. Both bachelor of science and master of science degrees are available. It generally takes four or five years to earn a bachelor's degree and two additional years to earn a master's degree. A master's degree is not required to be a packaging engineer, although many professionals pursue advanced degrees, particularly if they plan to specialize in a specific area or do research. Many students take their first job in packaging once they have earned a bachelor's degree, while other students earn a master's degree immediately upon completing their undergraduate studies.

Students interested in this field often structure their own programs. In college, if no major is offered in packaging engineering, students can choose a related discipline, such as mechanical, industrial, electrical, chemical, materials, or systems engineering. It is useful to take courses in graphic design, computer science, marketing, and management.

Students enrolled in a packaging engineering program will usually take the following courses during their first two years: algebra, trigonometry, calculus, chemistry, physics, accounting, economics, finance, and communications. During the remaining years, classes focus on core packaging subjects such as packaging materials, package development, packaging line machinery, and product protection and distribution. Elective classes include topics concentrating on packaging and the environment, packaging laws and regulation, and technical classes on specific materials. Graduate studies, or those classes necessary to earn a master's degree, include advanced classes in design, analysis, and materials and packaging processes.

Certification or Licensing

Special licensing is required for engineers whose work affects the safety of the public. Much of the work of packaging engineers, however, does not require a license even though their work affects such factors as food and drug spoilage, protection from hazardous materials, or protection from damage. Licensing laws vary from state to state, but, in general, states have similar requirements. They require that an engineer must be a graduate of an approved engineering school, have four years of engineering experience, and passes the state licensing examination. A state board of engineering examiners administers the licensing and registration of engineers.

Professional societies offer certification to engineers instead of licensing. Although certification is not required, many professional engineers obtain it to show that they have mastered specified requirements and have reached a certain level of expertise.

For those interested in working with the specialized field of military packaging technology, the U.S. Army offers courses in this field. Generally, a person earns a bachelor of science degree in packaging engineering before taking these specialized courses. The National Institute of Packaging, Handling, and Logistic Engineers has information about the field of military packaging.

Other Requirements

Packaging engineers should have the ability to solve problems and think analytically and creatively. They must work well with people, both as a leader and as a team player. They should also be able to write and speak well in order to deal effectively with other workers and customers, and in order to document procedures and policies. In addition, a packaging engineer should have the ability to manage projects and people.

Exploring

To get firsthand experience in the packaging industry, students can call local manufacturers to see how they handle and package their products. Often, factories will allow visitors to tour their manufacturing and packaging facilities.

Another way to learn about packaging is by observing the packaging that we encounter every day, such as containers for food, beverages, cosmetics, and household goods. Visit stores to see how products are packaged, stored, or displayed. Notice the shape and labeling on the container, its ease of use, durability for storage, convenience of opening and closing, disposability, and attractiveness.

Students may also explore their aptitude and interest in a packaging career through graphic design courses, art classes that include construction activities, and computer-aided design classes. Participating in hobbies that include designing and constructing objects from different types of materials can also be beneficial. Students may also learn about the industry by reading trade publications, such as *Packaging* and *Packaging Digest.*

Employers

Packaging engineers are employed by almost every manufacturing industry. Pharmaceutical, beverage, cosmetics, and food industries are major employers of packaging engineers. Some packaging engineers are hired to design and develop packaging while others oversee the actual production of the packages. Some companies have their own packaging facilities while other companies subcontract the packaging to specialized packing firms. Manufacturing and packaging companies can be large, multinational enterprises who manufacture, package, and distribute numerous products or they can be small operations that are limited to the production of one or two specific products. Specialized packaging companies hire employees for all aspects of the packaging design and production process. Worldwide manufacturing offers career opportunities around the world. The federal government and the armed services also have employment opportunities for packaging engineers.

Starting Out

College graduates with a degree in packaging or a related field of engineering should find it easy to get jobs as the packaging industry continues its rapid growth. Many companies send recruiters to college campuses to meet with graduating students and interview them for positions with their companies. Students can also learn about employment possibilities through their schools' placement services, job fairs, classified advertisements in newspapers and trade publications, and referrals from teachers. Students who have participated in an internship or work-study program through a college may learn about employment opportunities through contacts with industry professionals.

Students can also research companies they are interested in working for and apply directly to the person in charge of packaging or the personnel office.

Advancement

Beginning packaging engineers generally do routine work under the supervision of experienced engineers and may also receive some formal training through their company. As they become more experienced, they are given more difficult tasks and more independence in solving problems, developing designs, or making decisions.

Some companies provide structured programs in which packaging engineers advance through a sequence of positions to more advanced packaging engineering positions. For example, an entry-level engineer might start out by producing engineering layouts to assist product designers, advance to a product designer, and ultimately move into a management position.

Packaging engineers may advance from being a member of a team to a project supervisor or department manager. Qualified packaging engineers may advance through their department to become a manager or vice president of their company. To advance to a management position, the packaging engineer must demonstrate good technical and production skills and managerial ability. After years of experience, a packaging engineer might wish to become self-employed as a packaging consultant.

To improve chances for advancement, the packaging engineer may wish to get a master's degree in another branch of engineering or in business administration. Many executives in government and industry began their careers as engineers. Some engineers become patent attorneys by combining a law degree with their technical and scientific knowledge.

Many companies encourage continuing education throughout one's career and provide training opportunities in the form of in-house seminars and outside workshops. Taking advantage of any training offered helps one to develop new skills and learn technical information that can increase chances for advancement. Many companies also encourage their employees to participate in professional association activities. Membership and involvement in professional associations are valuable ways to stay current on new trends within the industry, to familiarize oneself with what other companies are doing, and to make contacts with other professionals in the industry. Many times, professionals learn about opportunities for advancement in new areas or at different companies through the contacts they have made at association events.

Earnings

Currently, the average starting salary for a packaging engineer with a bachelor's degree is about $35,000 per year. The mid-range salary is $45,000, with packaging engineers easily earning $60,000 or more as they gain experience and advance within a company.

Benefits vary from company to company but can include any of the following: medical, dental, and life insurance; paid vacations, holidays, and sick days; profit sharing; 401(k) plans; bonus and retirement plans; and educational assistance programs. Some employers pay fees and expenses for participation in professional associations.

Work Environment

The working conditions for packaging engineers vary with the employer and with the tasks of the engineer. Those who work for companies that make packaging materials or who direct packaging operations might work around noisy machinery. Generally, they have offices near the packaging operations where they consult with others in their department, such as packaging machinery technicians and other engineers.

Packaging engineers also work with nontechnical staff such as designers, artists, and marketing and financial people. Packaging engineers must be alert to keeping up with new trends in marketing and technological developments.

Most packaging engineers have a five-day, 40-hour workweek, although overtime is not unusual. In some companies, particularly during research and design stages, product development, and the start up of new methods or equipment, packaging engineers may work 10-hour days or longer and work on weekends.

Some travel may be involved, especially if the packaging engineer is also involved in sales. Also, travel between plants may be necessary to coordinate packaging operations. At various stages of developing packaging, the packaging engineer will probably be engaged in hands-on activities. These activities may involve handling objects, working with machinery, carrying light loads, and using a variety of tools, machines, and instruments.

The work of packaging engineers also involves other, social concerns such as consumer protection, environmental pollution, and conservation of natural resources. Packaging engineers are constantly searching for safer, tamper-proof packaging, especially because harmful substances have been found in some food, cosmetics, and drugs. They also experiment with new packaging materials and utilize techniques to conserve resources and reduce the disposal problem. Many environmentalists are concerned with managing the waste from discarded packages. Efforts are being made to stop littering; to recycle bottles, cans, and other containers; and to use more biodegradable substances in packaging materials. The qualified packaging engineer, then, will have a broad awareness of social issues.

Outlook

The packaging industry, which employs more than a million people, offers almost unlimited opportunities for packaging engineers. Packaging engineers work in almost any industry because virtually all manufactured products need one or more kinds of packaging. Some of the industries with the fastest growing packaging needs are food, drugs, and cosmetics.

The demand for packaging engineers is expected to increase faster than the average for all occupations as newer, faster ways of packaging are continually being sought to meet the needs of economic growth, world trade expansion, and the environment. Increased efforts are also being made to develop packaging that is easy to open for the growing aging population and those persons with disabilities.

For More Information

The following sources can provide information on educational programs and the packaging industry.

Institute of Electrical and Electronics Engineers (IEEE)
1828 L Street NW, Suite 1202
Washington, DC 20036-5104
Tel: 202-785-0017
Web: http://www.ieee.org/usab

Institute of Packaging Professionals
481 Carlisle Drive
Herndon, VA 22070-4823
Tel: 703-318-8970
Web: http://www.pakinfo-world.org/iopp/

National Institute of Packaging, Handling, and Logistic Engineers
6902 Lyle Street
Lanham, MD 20706
Tel: 301-459-9105

Packaging Education Forum
481 Carlisle Drive
Herndon, VA 22070-4823
Tel: 703-318-8970
Web: http://www.pakinfo-world.org

Packaging Machinery Manufacturers Institute
4350 North Fairfax Drive, Suite 600
Arlington, VA 22203
Tel: 703-243-8555
Web: http://www.packexpo.com

Petroleum Engineers

	School Subjects
Mathematics Physics	

	Personal Skills
Helping/teaching Technical/scientific	

	Work Environment
Indoors and outdoors One location with some travel	

	Minimum Education Level
Bachelor's degree	

	Salary Range
$42,000 to $53,000 to $67,000	

	Certification or Licensing
Recommended	

	Outlook
Decline	

Overview

Petroleum engineers apply the principles of geology, physics, and the engineering sciences to the recovery, development, and processing of petroleum. As soon as an exploration team has located an area that could contain oil or gas, petroleum engineers begin their work, which includes determining the best location for drilling new wells, as well as the economic feasibility of developing them. They are also involved in operating oil and gas facilities, monitoring and forecasting reservoir performance, and utilizing enhanced oil recovery techniques that extend the life of wells.

History

Within a broad perspective, the history of petroleum engineering can be traced back hundreds of millions of years to when the remains of plants and animals blended with sand and mud and transformed into rock. It is from this ancient underground rock that petroleum is taken, for the organic matter of the plants and animals decomposed into oil during these millions of years and accumulated into pools deep underground.

In primitive times, people did not know how to drill for oil; instead, they collected the liquid substance after it had seeped to above ground surfaces. Petroleum is known to have been used at that time for caulking ships and for concocting medicines.

Petroleum engineering as we know it today was not established until the mid-1800s, an incredibly long time after the fundamental ingredients of petroleum were deposited within the earth. In 1859, the American Edwin Drake was the first person to ever pump the so-called rock oil from under the ground, an endeavor that, before its success, was laughed at and considered impossible. Forward-thinking investors, however, had believed in the operation and thought that underground oil could be used as inexpensive fluid for lighting lamps and for lubricating machines (and therefore could make them rich). The drilling of that first well, in Titusville, Pennsylvania (1869), ushered in a new worldwide era—the oil age.

At the turn of the century, petroleum was being distilled into kerosene, lubricants, and wax. Gasoline was considered a useless by-product and was run off into rivers as waste. However, this changed with the invention of the internal combustion engine and the automobile. By 1915 there were more than half a million cars in the United States, virtually all of them powered by gasoline.

Edwin Drake's drilling operation struck oil 70 feet below the ground. Since that time, technological advances have been made, and a professional field of petroleum engineering has been established. Today's operations drill as far down as six miles. Because the United States began to rely so much on oil, the country contributed significantly to creating schools and educational programs in this engineering discipline. The world's first petroleum engineering curriculum was devised in the United States in 1914; today there are 30 U.S. universities that offer petroleum engineering degrees.

The first schools were concerned mainly with developing effective methods of locating oil sites and with devising efficient machinery for drilling wells. Over the years, as sites have been depleted, engineers have been more concerned with formulating methods for extracting as much oil as possible from each well. Today's petroleum engineers focus on issues such as computerized drilling operations; however, because usually only about 40 to 60 percent of each site's oil is extracted, engineers must still deal with designing optimal conditions for maximum oil recovery.

The Job

Petroleum engineer is a rather generalized title that encompasses several specialties, each one playing an important role in ensuring the safe and productive recovery of oil and natural gas. In general, petroleum engineers are involved in the entire process of oil recovery, from preliminary steps such as analyzing cost factors to the last stages such as monitoring the production rate and then repacking the well after it has been depleted.

Petroleum engineering is closely related to the separate engineering discipline of geoscience engineering. Before petroleum engineers can begin work on an oil reservoir, prospective sites must first be sought by geological engineers, along with geologists and geophysicists. These scientists determine whether a site has potential oil. Petroleum engineers develop plans for drilling. Drilling is usually unsuccessful, with eight out of 10 test wells being "dusters" (dry wells) and only one of the remaining two test wells having enough oil to be commercially producible. When a significant amount of oil is discovered, engineers can begin their work of maximizing oil production at the site. The development company's engineering manager oversees the activities of the various petroleum engineering specialties, including reservoir engineers, drilling engineers, and production engineers.

Reservoir engineers use the data gathered by the previous geoscience studies and estimate the actual amount of oil that will be extracted from the reservoir. It is the reservoir engineers who determine whether the oil will be taken by primary methods (simply pumping the oil from the field) or by enhanced methods (using additional energy such as water pressure to force the oil up). The reservoir engineer is responsible for calculating the cost of the recovery process relative to the expected value of the oil produced, and simulates future performance using sophisticated computer models. Besides performing studies of existing company-owned oil fields, reservoir engineers also evaluate fields the company is thinking of buying.

Drilling engineers work with geologists and drilling contractors to design and supervise drilling operations. They are the engineers involved with the actual drilling of the well. They ask, What will be the best methods for penetrating the earth? It is the responsibility of these workers to supervise the building of the derrick (a platform, constructed over the well, that holds the hoisting devices), choose the equipment, and plan the drilling methods. Drilling engineers must have a thorough understanding of the geological sciences so that they can know, for instance, how much stress to place on the rock being drilled.

Production engineers determine the most efficient methods and equipment to optimize oil and gas production. For example, they establish proper pumping unit configuration and perform tests to determine well fluid levels and pumping load. They plan field workovers and well stimulation techniques such as secondary and tertiary recovery (for example, injecting steam, water, or a special recovery fluid) to maximize field production.

Various research personnel are involved in this field; some are more specialized than others. They include the *research chief engineer*, who directs studies related to the design of new drilling and production methods, and the *oil-well equipment research engineer*, who directs research to design improvements in oil-well machinery and devices; and the *oil-field equipment test engineer*, who conducts experiments to determine the effectiveness and safety of these improvements.

In addition to all of the above, sales personnel play an important part in the petroleum industry. *Oil-well equipment and services sales engineers* sell various types of equipment and devices used in all stages of oil recovery. They provide technical support and service to their clients, including oil companies and drilling contractors.

Requirements

High School

High school students can prepare for college engineering programs by taking courses in mathematics, physics, chemistry, geology, and computer science; economics, history, and English are also highly recommended because these subjects improve communications and management skills. Mechanical drawing and foreign languages will also be helpful. Students should try for courses that are taught at the honors level.

Postsecondary Training

A bachelor's degree in engineering is the minimum requirement. In college, one can follow either a specific petroleum engineering curriculum or a program in a closely related field, such as geophysics or mining engineering. In the United States, there are about 30 universities and colleges that offer programs that concentrate on petroleum engineering; many of these are located in California and Texas. The first two years toward the bachelor of science degree involve the study of many of the same subjects taken in high school, only at an advanced level, as well as basic engineering courses. In the junior and senior years, students take more specialized courses: geology, formation evaluation, properties of reservoir rocks and fluids, well drilling, properties of reservoir fluids, petroleum production, and reservoir analysis.

Because the technology changes so rapidly, many petroleum engineers continue their education to receive a master's degree and then a doctorate. Both command higher salaries and often result in better advancement opportunities. Those who work in research and teaching positions usually need these higher credentials.

Students considering an engineering career in the petroleum industry should be aware that the industry uses all kinds of engineers. Those with chemical, electrical, geoscience, mechanical, environmental, and other engineering degrees are also employed in this field.

Certification or Licensing

Many jobs require that the engineer be licensed as a Professional Engineer (P.E.), which is often needed on certain public projects. To be licensed, candidates must have a degree from an engineering program accredited by the Accreditation Board for Engineering and Technology. Additional requirements for obtaining the P.E. license vary from state to state, but all applicants must take an exam and have several years of related experience on the job or in teaching.

Other Requirements

Students thinking about this career need to enjoy science and math. You need to be creative problem-solvers who like to come up with new ways to get things done and try them out. They need to be curious, wanting to know why and how things are done. You also need to be logical thinkers with a capacity for detail, and you must be good communicators who can work well with others.

Exploring

One of the most satisfying ways to explore this occupation is to participate in Junior Engineering Technical Society (JETS) programs. JETS participants enter engineering design and problem-solving contests and learn team development skills, often with an engineering mentor. Science fairs and clubs also offer fun and challenging ways to learn about engineering.

Certain students are able to attend summer programs held at colleges and universities that focus on material not traditionally offered in high school. Usually these programs include recreational activities such as basketball, swimming, and track and field. For example, Worcester Polytechnic Institute offers the Frontiers program, a 13-day residential session for high school seniors. The American Indian Science and Engineering Society (AISES) also sponsors two- to six-week mathematics and science camps that are open to American Indian students and held at various college campuses.

Talking with someone who has worked as a petroleum engineer would also be a very helpful and inexpensive way of exploring this field. One good way to find an experienced person to talk to is through Internet sites that feature career areas to explore, industry message boards and mailing lists.

Other ways to explore this career include possible tours of oilfields or corporate sites (contact the public relations department of oil companies) and summer and other temporary jobs in the petroleum industry on drilling and production crews. Trade journals, high school guidance counselors, the placement office at technical or community colleges, and the associations listed at the end of this article are other helpful resources.

Employers

Petroleum engineers are employed by major oil companies, as well as smaller oil companies. They work in oil exploration and production. Some petroleum engineers are employed by consulting companies and equipment suppliers. The government is also an employer of engineers.

Starting Out

The most common and perhaps the most successful way to obtain a petroleum engineering job is to apply with the student placement services department at the college that you attend. Oil companies often have recruiters who seek potential graduates while they are in their last year of engineering school.

Applicants are also advised to simply check the job sections of major newspapers and apply directly to companies seeking employees; they should also keep informed of the general national employment outlook in this industry by reading trade and association journals, such as the Society of Petroleum Engineers' *Journal of Petroleum Technology*.

Engineering internships or co-op programs where students attend classes for a portion of the year and then work in an engineering-related job for the remainder of the year allow students to graduate with valuable work experience sought by employers. Many times these students are employed full time after graduation at the place where they had their internship or co-op job.

As in most engineering professions, entry-level petroleum engineers first work under the supervision of experienced professionals for a number of years. New engineers usually are assigned to a field location where they learn different aspects of field petroleum engineering. Initial responsibilities may include well productivity, reservoir and enhanced recovery studies, production equipment and application design, efficiency analyses, and economic evaluations. Field assignments are followed by other opportunities in regional and headquarters offices.

Advancement

After several years working under professional supervision, engineers can begin to move up to higher levels. Workers often formulate a choice of direction during their first years on the job. In the operations division, petroleum engineers can work their way up from the field to district, division, and then operations manager. Some engineers work through various engineering positions from field engineer to staff, then division, and finally chief engineer on a project. Some engineers may advance into top executive management. In any position, however, continued enrollment in educational courses is usually required to keep abreast of technological progress and changes. After about four years of work experience, engineers tend to apply for a P.E. license so they can be certified to work on a larger number of projects.

Others get their master's or doctoral degree so they can advance to more prestigious research engineering, university-level teaching, or consulting positions. Also, there are opportunities for petroleum engineers to transfer to many other occupations, such as economics, environmental management, and groundwater hydrology. Finally, there are the workers with entrepreneurial spirit who become independent operators of their own oil companies.

Earnings

Salaries for entry-level engineers with a bachelor's degree are often higher than for workers in any other field. Furthermore, petroleum engineers tend to make the highest starting salaries of all engineers. According to the *American Almanac of Jobs and Salaries,* entry level petroleum engineers earn an average of $42,000; those with some experience earn an average of $53,000; and those with extensive experience earn an average of $67,000.

Salary increases tend to reflect changes in the petroleum industry as a whole. When the price of oil is high, salaries can be expected to grow; low oil prices often result in stagnant wages.

Fringe benefits are good. Most employers provide health and accident insurance, sick pay, retirement plans, profit-sharing plans, and paid vacations. Education benefits are also competitive.

Work Environment

Petroleum engineers work all over the world—the high seas, remote jungles, vast deserts, plains, and mountain ranges. In the United States, oil or natural gas is produced in 33 states, with most sites located in Texas, Alaska, Louisiana, California, and Oklahoma, plus offshore regions. Many other U.S. engineers work in other oil-producing areas such as the Arctic Circle, China's Tarim Basin, and Saudi Arabia. Assignments to remote foreign locations can make family life difficult. Those working overseas may live in company-supplied housing.

Some petroleum engineers, such as drilling engineers, work primarily out in the field at or near drilling sites in all kinds of weather and environments. The work can be dirty and dangerous. Responsibilities such as making reports, conducting studies of data, and analyzing costs are usually tended in offices either away from the site or in temporary work trailers.

Other engineers work in offices in cities of varying sizes, with only occasional visits to an oil field. Research engineers work in laboratories much of the time, while those who work as professors spend most of their time on campuses and at other teaching areas. Workers involved in economics, management, consulting, and government service tend to spend their work time exclusively in indoor offices.

Outlook

The opportunity for employment in this field directly depends on the world price for oil and gas. Conditions at the beginning of 1996 included a worldwide surplus of oil, which is expected to continue. At the same time, domestic conservation of oil by industry and the public has reduced the demand for oil. The surplus has resulted in low oil prices. For these reasons, the number of job openings for petroleum engineers is expected to decline through 2006. Even so, some opportunities for petroleum engineers will exist because the number of degrees granted in petroleum engineering is low.

The challenge for petroleum engineers in the past decade has been to develop technology that lets drilling and production be economically feasible even in the face of low oil prices. For example, engineers had to rethink how they worked in deep water. They used to believe deep wells would collapse if too much oil was pumped out at once. But the high costs of working in deep water plus low oil prices made low volumes uneconomical. So engineers learned how to boost oil flow by slowly upping the quantities wells pumped by improving valves, pipes, and other equipment used. Engineers have also cut the cost of deep-water oil and gas production in the Gulf of Mexico, predicted to be one of the most significant exploration hot spots in the world for the next decade, by placing wellheads on the ocean floor instead of on above-sea production platforms.

Cost-effective technology that permits new drilling and increases production will continue to be essential in the profitability of the oil industry. Therefore, petroleum engineers will continue to have a vital role to play, even in this age of streamlined operations and company restructurings.

For More Information

This trade association represents employees in the petroleum industry. Free videos, fact sheets, and informational booklets are available to educators.

American Petroleum Institute
1220 L Street, NW
Washington, DC 20005
Tel: 202-682-8000
Web: http://www.api.org

For a petroleum engineering career brochure, a list of petroleum engineering schools, and scholarship information, contact:

Society of Petroleum Engineers
PO Box 833836
Richardson, TX 75083-3836
Tel: 972-952-9393
Web: http://www.spe.org

For information on careers in geology, write:

American Association of Petroleum Geologists
1444 S. Boulder Avenue
Tulsa, OK 74119
Tel: 918-584-2555
Web: http://www.aapg.org

For information about JETS programs, products, and engineering career brochures (all disciplines), contact:

Junior Engineering Technical Society, Inc.
1420 King Street, Suite 405
Alexandria, VA 22314-2715
Tel: 703-548-5387
Email: jets@nae.edu
Web: http://www.jets.org

For career advice and booklets on engineering and geoscience careers, contact:

The Association of Professional Engineers, Geologists and Geophysicists of Alberta
15th Floor, Tower One, Scotia Place
10060 Jasper Avenue
Edmonton, AB T5J 4A2 Canada
Tel: 800-661-7020
Web: http://www.apegga.org

For information on AISES precollege programs and scholarships, contact:

American Indian Science & Engineering Society
5655 Airport Road
Boulder, CO 80301
Tel: 303-939-0023

For Opportunities for Performance, *a booklet describing professional jobs at Phillips Petroleum Company, contact:*

Phillips Petroleum Company, Employment & College Relations
180 Plaza Office Building
Bartlesville, OK 74004
Tel: 918-661-6385

For Oil, *a booklet describing all phases of the oil industry, and information on careers with Shell, contact:*

Shell Oil Company
External Affairs
PO Box 2463
Houston, TX 77252-2463
Tel: 713-241-6161
Web: http://www.shell.com

Plastics Engineers

Chemistry Computer science	School Subjects
Mechanical/manipulative Technical/scientific	Personal Skills
Primarily indoors Primarily one location	Work Environment
Bachelor's degree	Minimum Education Level
$30,000 to $46,500 to $87,750	Salary Range
Required for certain positions	Certification or Licensing
Faster than the average	Outlook

Overview

Plastics engineers engage in the manufacture, fabrication, and end use of existing materials, as well as with the development of new materials, processes, and equipment. The term, plastics engineer, encompasses a wide variety of applications and manufacturing processes. Depending on the processes involved, plastics engineers develop everything from the initial part design to the processes and automation required to produce and finish the production parts.

History

Thermoplastics, plastics that soften with heat and harden when cooled, were discovered in France in 1828. In the United States in 1869, a printer, John Wesley Hyatt, created celluloid in the process of attempting to create an alternate material to supplement ivory in billiard balls. His invention, patented in 1872, brought about a revolution in production and manufacturing. By 1892, over 2,500 articles were being produced from celluloid. Among these inventions were frames for eyeglasses, false teeth, the first movie film, and, of course, billiard balls. Celluloid did have its drawbacks. It could not be molded and it was highly flammable.

It was not until 1909 that the Belgian-American chemist Leo H. Baekeland (1863-1944) produced the first synthetic plastic. This product replaced natural rubber in electrical insulation and was used for phone hand-sets and automobile distributor caps and rotors, and is still used today. Other plastics materials have been developed steadily. The greatest variety of mate-rials and applications, however, came during World War II, when the war effort brought about a need for changes in clothing, consumer goods, trans-portation, and military equipment.

Today, plastics manufacturing is a major industry whose products play a vital role in many other industries and activities around the world. It is dif-ficult to find an area of our lives where plastic does not play some role. Plastics engineers apply their skills to a vast array of professional fields. For example, plastics engineers assisting those in the medical field may help to further develop artificial hearts, replacement limbs, artificial skin, implantable eye lenses, and specially designed equipment that will aid sur-geons and other health professionals in the operating room.

The Job

Plastics engineers perform a wide variety of duties depending on the type of company they work for and the products it produces. Plastics engineers, for example, may develop ways to produce clear, durable plastics to replace glass in areas where glass cannot be used. Others design and manufacture light-weight parts for aircraft and automobiles, or create new plastics to replace metallic or wood parts that have come to be too expensive or hard to obtain. Others may be employed to formulate less-expensive, fire-resistant plastics for use in the construction of houses, offices, and factories. Plastics engineers may also develop new types of biodegradable molecules that are friendly to the environment, reducing pollution and increasing recyclability.

Plastics engineers perform a variety of duties. Some of their specific job titles and duties include: *application engineers,* who develop new processes and materials in order to create a better finished product; *process engineers,* who oversee the production of reliable, high quality, standard materials; and *research specialists,* who use the basic building blocks of matter to discover and create new materials.

In the course of their day, plastics engineers must solve a wide variety of internal production problems. Duties include making sure the process is consistent to insure creation of accurate and precise parts and making sure parts are handled and packaged efficiently, properly, and cheaply. Each part is unique in this respect.

Computers are increasingly being used to assist in the production process. Plastics engineers use computers to calculate part weight and cycle times; for monitoring the process on each molding press; for designing parts and molds on the CAD system; for tracking processes and the labor in the mold shop; and to transfer engineering files over the Internet.

Plastics engineers also help customers solve problems that may emerge in part design—finding ways to make a part more moldable or to address possible failures or inconsistencies in the final design. Factors that may make a part difficult to mold include: thin walls, functional or cosmetic factors, sections that are improperly designed that will not allow the part to be processed efficiently, or inappropriate material selection which results in an improperly created part.

Plastics engineers also coordinate mold-building schedules and activities with tool vendors. Mold-building schedules consist of the various phases of constructing a mold, from the development of the tool and buying of materials (and facilitating their timely delivery), to estimating the roughing and finishing operations. Molds differ depending on the size of the tool or product, the complexity of the work orders, and the materials required to build the mold.

Most importantly, plastics engineers must take an application that is difficult to produce and make it (in the short period of time allowed) profitable to their company, while still satisfying the needs of the customer.

Requirements

High School

Follow your school's college prep program by taking classes in English, government, foreign language, and history. You should take additional classes in mathematics and the sciences, particularly chemistry and physics. Computer classes are also important. You should also take voc-tech, drafting, and other classes that involve you directly with design and manufacture.

Postsecondary Training

The level of education required beyond high school varies greatly depending on the types of plastics processes involved. Most plastic companies do not require a bachelor's degree in plastics engineering. Companies that design proprietary parts usually require a bachelor's or advanced degree in mechanical engineering. The field of plastics engineering, overall, is still a field where people with the proper experience are scarce—experience is a key factor in qualifying a person for an engineering position.

To pursue an associate's or bachelor's degree in plastics engineering, you should contact the Society of the Plastics Industry (SPI) or the Society of Plastics Engineers (SPE) for information about two-year and four-year programs. Plastics programs are sometimes listed under polymer science, polymer engineering, materials science, and materials engineering. The Society of Plastics Engineers offers scholarships to some students enrolled in engineering programs. Awards range up to $4,000 annually, and are renewable for up to three additional years. Certain branches of the military also provide training in plastics engineering.

Certification or Licensing

Some states may require that engineers be licensed. Though national certification isn't required, SPE has established The Institute for Plastics Certification. To receive this certification, technologists and engineers with the required amount of education and experience can receive certification after passing an exam.

Other Requirements

You need to have good mechanical aptitude, in order to develop the plastics parts and the tooling necessary to develop these parts. You must have thorough knowledge of the properties of plastic and of the processes which occur. There are thousands of different materials which you may encounter in the course of your workday. You also must be imaginative and creative in order to be able to solve any problems which might arise from new applications or in the transition/transformation of a mechanical metal part to that of a plastic one.

Exploring

You can gain some insight into plastics careers by looking at the industry publications *Plastics Engineering* and *Plastics Industry News*. SPI also publishes a careers brochure. *Opportunities in Plastics Careers,* by Jan Bone, is a useful overview of the plastics industry. There are numerous chapters which focus on job-finding skills and financial aid opportunities.

High school students may seek to join JETS (Junior Engineering Technical Society), a program which provides organized engineering-related activities. Students, through group activities, can gain practice in problem solving, scientific reasoning, and actual real life experience with the real world of engineering.

A high school counselor, science, or shop teacher may be able to arrange a presentation or question-and-answer session with a plastics engineer, or even a tour of a local plastics manufacturer. There are also student chapters of SPI and SPE which provide opportunities to gain valuable experience and contacts with similarly interested people. You may also be able to find a summer job at a plastics-processing plant to learn the basics and experience the varied areas involved with producing plastics parts.

Employers

Plastics engineers work for the manufacturers of plastic products, materials, and resins. Major plastics employers in the United States include DuPont, General Motors, and Owens-Corning. Some of the top thermoforming companies are in Illinois: Tenneco Packaging, Solo Cup Company, and Ivex Packaging Corporation are a few of them. Michigan has some of the top injection molding companies, including Textron, Lear Corporation, UT Automotive, and Venture Industries Corporation. But large plastics companies are located all across the country. According to the SPI, the top plastics industry states ranked by employment are California, Ohio, Michigan, Illinois, and Texas.

Starting Out

To get a job as a plastics engineer, you'll need considerable experience in the plastics industry or a college degree. A variety of starting points exist within the industry. Experienced plastics setup and process technicians can use their skills to advance to engineering responsibilities. Many plastics engineers start out as tool and die makers or moldmakers before they move into engineering positions.

For those who receive their plastics knowledge through advanced education, jobs can be obtained through the placement programs of their universities and technical schools. Also, many major companies recruit plastics engineers on college campuses. SPE's Web site features the "online plastics employment network," a database of job openings.

Advancement

The advanced training, expertise, and knowledge of experienced plastics engineers allows them the luxury of migrating to almost any position within the plastics industry. Engineers may also advance to supervisory or management positions, for example, becoming director of engineering for their entire plant or division. Further advancement may come in the form of employment at larger companies.

Experienced plastics engineers, as a result of their expertise in materials and matching products to applications, are good candidates for sales and marketing jobs. They may also train the engineers of tomorrow by becoming teachers at technical schools or colleges or by writing for a technical trade journal.

Earnings

According to a survey by the Institute of Industrial Engineers, the median annual salary for engineers working for rubber and plastics products manufacturers is $46,500. Those starting out may make less than $30,000, but those with more education and experience can make well over $50,000. In a 1998 industry report, the American Association of Engineering Societies stated that engineers with 10 years experience in the plastics industry made an average of $64,000 a year. Those with 14 years experience made $69,800, and those with 25 years experience made $87,750.

Benefits for plastics engineers usually include paid vacations and sick days, pension plans, and health and dental insurance. Depending on the size of the company, engineers may be offered production bonuses, stock options, and paid continuing education.

Work Environment

Plastics engineers are constantly busy as they deal with people at all levels and phases of the manufacturing process. Dress codes may be formal since plastics engineers interact with customers frequently during the course of a day. Engineers may be required to work more than a standard eight-hour day and also some Saturdays when a specific project is on a deadline.

As a plastics engineer, you may work directly with design materials in a laboratory, or sit at a computer in an office. You may spend some hours working alone, as well some hours working as part of a team. You may only be involved in certain aspects of a project, or you may work on a project from the original design to final testing of a product.

Outlook

The future of plastics engineering is very bright. Three or four new plastics materials are being discovered every day. Most industries are less likely to lay off plastics engineers than other types of workers. More industries are incorporating plastics into their product lines, which will create more opportunities for qualified plastics engineers. As more plastics products are substituted for glass, paper, and metal products and parts, plastics engineers will be needed to oversee design and production processes. An example of this change is in the automotive industry, where a high percentage of engine parts will eventually be made of plastic. Plastics engineers will increasingly be required to develop environmentally friendly products and processes, and play a role in developing easily recyclable products for certain industries.

Many opportunities exist in smaller companies, such as plastics parts suppliers. Many openings will come as a result of experienced engineers who advance to sales, management, or other related occupations within the plastics industry. Those with the most advanced skills and experience, as always, will enjoy the best future career outlook.

For More Information

For information on obtaining a copy of Plastics Engineering *and information on college scholarships, contact:*

Society of Plastics Engineers
PO Box 403
Brookfield, CT 06804-0403
Tel: 203-775-0471
Web: http://www.4spe.org

For a career brochure and information about college programs, contact:

Society of the Plastics Industry
1801 K Street, NW, Suite 600K
Washington, DC 20006-1301
Tel: 202-974-5200
Web: http://www.socplas.org

For information on membership and programs, contact:

Junior Engineering Technical Society, Inc.
1420 King Street, Suite 405
Alexandria, VA 22314-2794
Tel: 703-548-5387
Email: jets@nae.edu
Web: http://www.jets.org

Quality Control Engineers and Technicians

Mathematics Physics	School Subjects
Mechanical/manipulative Technical/scientific	Personal Skills
Primarily indoors Primarily one location	Work Environment
Associate's degree	Minimum Education Level
$17,000 to $40,000 to $70,000	Salary Range
Voluntary	Certification or Licensing
About as fast as the average	Outlook

Overview

Quality control engineers plan and direct procedures and activities involved in the processing and production of materials and goods in order to ensure specified standards of quality. They select the best techniques for a specific process or method, determine the level of quality needed, and take the necessary action to maintain or improve quality performance. *Quality control technicians* assist quality control engineers in devising quality control procedures and methods, implement quality control techniques, test and inspect products during different phases of production, and compile and evaluate statistical data to monitor quality levels.

History

Quality control technology is an outgrowth of the Industrial Revolution. As it began in England in the 18th century, each person involved in the manufacturing process was responsible for a particular part of the process. The worker's responsibility was further specialized by the introduction of the concept of interchangeable parts in the late 18th and early 19th centuries. In a manufacturing process using this concept, a worker could concentrate on making just one component, while other workers concentrated on creating other components. Such specialization led to increased production efficiency, especially as manufacturing processes became mechanized during the early part of the 20th century. It also meant, however, that no one worker was responsible for the overall quality of the product. This led to the need for another kind of specialized production worker whose primary responsibility was not one aspect of the product but rather its overall quality.

This responsibility initially belonged to the mechanical engineers and technicians who developed the manufacturing systems, equipment, and procedures. After World War II, however, a new field emerged that was dedicated solely to quality control. Along with specially trained persons to test and inspect products coming off assembly lines, new instruments, equipment, and techniques were developed to measure and monitor specified standards.

At first, quality control engineers and technicians were primarily responsible for random checks of products to ensure they met all specifications. This usually entailed testing and inspecting either finished products or products at various stages of production.

During the 1980s, a quality movement spread across the United States. Faced with increased global competition, especially from Japanese manufacturers, many U.S. companies sought to improve quality and productivity. Quality improvement concepts, such as total quality management, continuous improvement, quality circles, and zero defects gained popularity and changed the way companies viewed quality and quality control practices. A new philosophy emerged, emphasizing quality as the concern of all individuals involved in producing goods and directing that quality be monitored at all stages of manufacturing—not just at the end of production or at random stages of manufacturing.

Today, most companies focus on improving quality during all stages of production, with an emphasis on preventing defects rather than merely identifying defective parts. There is an increased use of sophisticated automated equipment that can test and inspect products as they are manufactured. Automated equipment includes cameras, X rays, lasers, scanners, metal detectors, video inspection systems, electronic sensors, and machine vision systems that can detect the slightest flaw or variance from accepted tolerances. Many companies use statistical process control to record levels of quality and determine the best manufacturing and quality procedures. Quality control engineers and technicians work with employees from all departments of a company to train them in the best quality methods and to seek improvements to manufacturing processes to further improve quality levels.

Many companies today are seeking to conform to international standards for quality, such as ISO 9000, in order to compete with foreign companies and to sell products to companies in countries around the world. These standards are based on concepts of quality regarding industrial goods and services and include documenting quality methods and procedures.

The Job

Quality control engineers are responsible for developing, implementing, and directing processes and practices that result in the desired level of quality for manufactured parts. They identify standards to measure the quality of a part or product, analyze factors that affect quality, and determine the best practices to ensure quality.

Quality control engineers set up procedures to monitor and control quality, devise methods to improve quality, and analyze quality control methods for effectiveness, productivity, and cost factors. They are involved in all aspects of quality during a product's life cycle. Not only do they focus on ensuring quality during production operations, they also get involved in product design and product evaluation. Quality control engineers may be specialists who work with engineers and industrial designers during the design phase of a product, or they may work with sales and marketing professionals to evaluate reports from consumers on how well a product is performing. Quality control engineers are responsible for ensuring that all incoming materials used in a finished product meet required standards and that all instruments and automated equipment used to test and monitor parts during production perform properly. They supervise and direct workers involved in assuring quality, including quality control technicians, inspectors, and related production personnel.

Quality control technicians work with quality control engineers in designing, implementing, and maintaining quality systems. They test and inspect materials and products during all phases of production in order to ensure they meet specified levels of quality. They may test random samples of products or monitor production workers and automated equipment that inspect products during manufacturing. Using engineering blueprints, drawings, and specifications, they measure and inspect parts for dimensions, performance, and mechanical, electrical, and chemical properties. They establish *tolerances,* or acceptable deviations from engineering specifications, and direct manufacturing personnel in identifying rejects and items that need to be reworked. They monitor production processes to be sure that machinery and equipment are working properly and set to established specifications.

Quality control technicians also record and evaluate test data. Using statistical quality control procedures, technicians prepare charts and write summaries about how well a product conforms to existing standards. Most important, they offer suggestions to quality control engineers on how to modify existing quality standards and manufacturing procedures. This helps to achieve the optimum product quality from existing or proposed new equipment.

Quality control technicians may specialize in any of the following areas: product design, incoming materials, process control, product evaluation, inventory control, product reliability, research and development, and administrative applications. Nearly all industries employ quality control technicians.

Requirements

High School

In high school, prospective engineers and technicians should take classes in English, mathematics (including algebra and geometry), physical sciences, physics, and chemistry. They should also take shop, mechanical drawing, and computer courses. Students should especially seek English courses that will develop their reading skills, the ability to write short reports with good organization and logical development of ideas, and the ability to speak comfortably and effectively in front of a group.

Postsecondary Training

Quality control engineers must have a bachelor's degree in engineering. Many quality control engineers receive degrees in industrial or manufacturing engineering. Some receive degrees in metallurgical, mechanical, electrical, or chemical engineering depending on where they plan to work. College engineering programs vary based on the type of engineering program. Most programs take four to five years to complete and include courses in mathematics, physics, and chemistry. Other useful courses include statistics, logistics, business management, and technical writing.

Educational requirements for quality control technicians vary by industry. Most employers of quality control technicians prefer to hire applicants who have received some specialized training. A small number of positions for technicians require a bachelor of arts or science degree. In most cases, though, completion of a two-year technical program is sufficient. Students enrolled in such a program at a community college or technical school take courses in the physical sciences, mathematics, materials control, materials testing, and engineering-related subjects.

Certification or Licensing

Although there are no licensing or certification requirements designed specifically for quality control engineers or technicians, some may need to meet special requirements that apply only within the industry employing them. Many quality control engineers and technicians pursue voluntary certification to indicate that they have achieved a certain level of competency, either through education or work experience. Such certification is offered through professional associations, such as the American Society for Quality Control (ASQC), and requires passing an examination. Many employers value this certification and regard it as a demonstration of professionalism.

Other Requirements

Quality control engineers need scientific and mathematical aptitudes, strong interpersonal skills, and leadership abilities. Good judgment is also needed, as quality control engineers must weigh all the factors influencing quality and determine procedures that incorporate price, performance, and cost factors.

Quality control technicians should enjoy and do well in mathematics, science, and other technical subjects and should feel comfortable using the language and symbols of mathematics and science. They should have good eyesight and good manual skills, including the ability to use hand tools. They should be able to follow technical instructions and to make sound judgments about technical matters. Finally, they should have orderly minds and be able to maintain records, conduct inventories, and estimate quantities.

Exploring

Because quality control engineers and technicians work in a wide variety of settings, prospective engineers and technicians who want to learn more about quality control technology can consider a range of possibilities for experiencing or further exploring such work. Quality control activities are often directly involved with manufacturing processes. Students may be able to get part-time or summer jobs in manufacturing settings, even if not specifically in the quality control area. Although this type of work may consist of menial tasks, it does offer firsthand experience and demonstrates interest to future employers.

Quality control engineers and technicians work with scientific instruments; therefore, academic or industrial arts courses that introduce different kinds of scientific or technical equipment will be helpful, along with electrical and machine shop courses, mechanical drawing courses, and chemistry courses with lab sections. Joining a radio, computer, or science club is also a good way to gain experience and to engage in team-building and problem-solving activities. Active participation in clubs is a good way to learn skills that will benefit you when working with other professionals in manufacturing and industrial settings.

Employers

The majority of quality control engineers and technicians are employed in the manufacturing sector of the economy. Because engineers and technicians work in all areas of industry, their employers vary widely in size, product, location, and prestige.

Starting Out

Students enrolled in two-year technical schools may learn of openings for quality control technicians through their schools' job placement services. Recruiters often visit these schools and interview graduating students for technical positions. Quality control engineers also may learn of job openings through their schools' job placement services, recruiters, and job fairs. In many cases, employers prefer to hire engineers who have some work experience in their particular industry. For this reason, applicants who have had summer or part-time employment or participated in a work-study or internship program have greater job opportunities.

Students may also learn about openings through help wanted ads or by using the services of state and private employment services. They also may apply directly to companies that employ quality control engineers and technicians. Students can identify and research such companies by using job resource guides and other reference materials available at most public libraries.

Advancement

Quality control technicians usually begin their work under the direct and constant supervision of an experienced technician or engineer. As they gain experience or additional education, they are given more responsible assignments. They can also become quality control engineers with additional education. Promotion usually depends upon additional training as well as job performance. Technicians who obtain additional training have greater chances for advancement opportunities.

Quality control engineers may have limited opportunities to advance within their companies. However, because quality control engineers work in all areas of industry, they have the opportunity to change jobs or companies to pursue more challenging or higher-paying positions. Quality control engineers who work in companies with large staffs of quality personnel can become quality control directors or advance to operations management positions.

Earnings

Earnings vary according to the type of work, the industry, and the geographical location. Quality control engineers earn salaries comparable to other engineers. Beginning engineers with a bachelor's degree generally earn between $31,000 and $35,000 a year. Those with master's degrees earn salaries of about $41,322 in their first jobs upon graduation. Experienced quality control engineers earn salaries ranging from $35,000 to $70,000.

Most beginning quality control technicians who are graduates of two-year technical programs earn salaries ranging from $17,000 to $21,000 a year. Experienced technicians with two-year degrees earn salaries that range from $21,000 to $36,000 a year; some senior technicians with special skills or experience may earn much more.

Most companies offer benefits that include paid vacations, paid holidays, and health insurance. Actual benefits depend upon the company, but may also include pension plans, profit sharing, 401(k) plans, and tuition assistance programs.

Work Environment

Quality control engineers and technicians work in a variety of settings, and their conditions of work vary accordingly. Most work in manufacturing plants, though the type of industry determines the actual environment. For example, quality control engineers in the metals industry usually work in foundries or iron and steel plants. Conditions are hot, dirty, and noisy. Other factories, such as for the electronics or pharmaceutical industries, are generally quiet and clean. Most engineers and technicians have offices separate from the production floor, but they still need to spend a fair amount of time there. Engineers and technicians involved with testing and product analysis work in comfortable surroundings, such as a laboratory or workshop. Even in these settings, however, they may be exposed to unpleasant fumes and toxic chemicals. In general, quality control engineers and technicians work inside and are expected to do some light lifting and carrying (usually not more than 20 pounds). Because many manufacturing plants operate 24 hours a day, some quality control technicians may need to work second or third shifts.

As with most engineering and technical positions, the work can be both challenging and routine. Engineers and technicians can expect to find some tasks repetitious and tedious. In most cases, though, the work provides variety and satisfaction from using highly developed skills and technical expertise.

Outlook

The employment outlook depends, to some degree, on general economic conditions. Although many economists forecast low to moderate growth in manufacturing operations through the year 2007, employment opportunities for quality control personnel should remain steady or slightly increase as many companies place increased emphasis on quality control activities.

Many companies are making vigorous efforts to make their manufacturing processes more efficient, lower costs, and improve productivity and quality. Opportunities for quality control engineers and technicians should be good in the food and beverage industries, pharmaceutical firms, electronics companies, and chemical companies. Quality control engineers and technicians also may find employment in industries using robotics equipment or in the aerospace, biomedical, bioengineering, environmental controls, and transportation industries. Lowered rates of manufacturing in the automotive and defense industries will decrease the number of quality control personnel needed for these areas. Declines in employment in some industries may occur because of the increased use of automated equipment that tests and inspects parts during production operations.

For More Information

American Society for Quality Control
PO Box 3005
Milwaukee, WI 53201-3005
Tel: 800-248-1946

Robotics Engineers and Technicians

Computer science Mathematics	School Subjects
Mechanical/manipulative Technical/scientific	Personal Skills
Primarily indoors Primarily one location	Work Environment
Bachelor's degree	Minimum Education Level
$20,000 to $35,000 to $60,000	Salary Range
None available	Certification or Licensing
About as fast as the average	Outlook

Overview

Robotics engineers design, develop, build, and program robots and robotic devices, including peripheral equipment and computer software used to control robots. *Robotics technicians* assist robotics engineers in a wide variety of tasks relating to the design, development, production, testing, operation, repair, and maintenance of robots and robotic devices.

History

Robots are devices that perform tasks ordinarily performed by humans; they seem to operate with an almost-human intelligence. The idea of robots can be traced back to the ancient Greek and Egyptian civilizations. An inventor from the first century AD, Hero of Alexandria, invented a machine that would automatically open the doors of a temple when the priest lit a fire in the altar.

During the later periods of the Middle Ages, the Renaissance, and the 17th and 18th centuries, interest in robot-like mechanisms turned mostly to automatons, devices that imitate human and animal appearance and activity but perform no useful task.

The Industrial Revolution inspired the invention of many different kinds of automatic machinery. One of the most important robotics inventions occurred in 1804: Joseph-Marie Jacquard's method for controlling machinery by means of a programmed set of instructions recorded on a punched paper tape that was fed into a machine to direct its movements.

The word *robot* and the concepts associated with it were first introduced in the early 1920s. They made their appearance in a play titled *R.U.R.,* which stands for Rossum's Universal Robots, written by Czechoslovakian dramatist Karel Capek. The play involves human-like robotic machines created to perform manual tasks for their human masters.

During the 1950s and 1960s, advances in the fields of automation and computer science led to the development of experimental robots that could imitate a wide range of human activity, including self-regulated and self-propelled movement (either on wheels or on legs), the ability to sense and manipulate objects, and the ability to select a course of action on the basis of conditions around them.

In 1954, George Devol designed the first programmable robot in the United States. He named it the Universal Automation, which was later shortened to Unimation, which also became the name of the first robot company. Hydraulic robots, controlled by numerical control programming, were developed in the 1960s and were used initially by the automobile industry in assembly line operations. By 1973, robots were being built with electric power and electronic controls, which allowed greater flexibility and increased uses.

Robotic technology has evolved significantly in the past few decades. Early robotic equipment, often referred to as first-generation robots, were simple mechanical arms or devices that could perform precise, repetitive motions at high speeds. They contained no artificial intelligence capabilities. Second-generation robots, which came into use in the 1980s, are controlled by minicomputers and programmed by computer language. They contain sensors, such as vision systems and pressure, proximity, and tactile sensors, which provide information about the outside environment. Third-generation robots, also controlled by minicomputers and equipped with sensory devices, are currently being developed. Referred to as "smart" robots, they can work on their own without supervision by an external computer or human being.

The evolution of robots is closely tied to the study of human anatomy and movement of the human body. The early robots were modeled after arms, then wrists. Second-generation robots include features that model human hands. Third-generation robots are being developed with legs and complex joint technology. They also incorporate multisensory input controls, such as ultrasonic sensors or sensors that can "sniff" and "taste."

The Job

The majority of robotics engineers and technicians work within the field of computer-integrated manufacturing or programmable automation. Using computer science technology, engineers design and develop robots and other automated equipment, including computer software used to program robots.

The title *robotics engineer* may be used to refer to any engineer who works primarily with robots. In many cases, these engineers may have been trained as mechanical, electronic, computer, or manufacturing engineers. A small, but growing, number of engineers trained specially in robotics are graduating from colleges and universities with robotics engineering or closely related degrees.

Robotics engineers have a thorough understanding of robotic systems and equipment and know the different technologies available to create robots for specific applications. They have a strong foundation in computer systems and how computers are linked to robots. They also have an understanding of manufacturing production requirements and how robots can best be used in automated systems to achieve cost efficiency, productivity, and quality. Robotics engineers may analyze and evaluate a manufacturer's operating system to determine whether robots can be used efficiently instead of other automated equipment or humans.

Many other types of engineers are also involved in the design, development, fabrication, programming, and operation of robots. Following are brief descriptions of these types of engineers and how they relate to robotics.

Electrical and electronics engineers research, design, and develop the electrical systems used in robots and the power supply, if it is electrical. These engineers may specialize in areas such as integrated circuit theory, lasers, electronic sensors, optical components, and energy power systems.

Mechanical engineers are involved in the design, fabrication, and operation of the mechanical systems of a robot. These engineers need a strong working knowledge of mechanical components such as gripper mechanisms, bearings, gears, chains, belts, and actuators. Some robots are controlled by pneumatic or mechanical power supplies and these engineers need to be specialists in designing these systems. Mechanical engineers also select the material used to make robots. They test robots once they are constructed.

Computer engineers design the computer systems that are used to program robots. Sometimes these systems are built into a robot and other times they are a part of separate equipment that is used to control robots. Some computer engineers also write computer programs.

Industrial engineers are specialists in manufacturing operations. They determine the physical layout of a factory to best utilize production equipment. They may determine the placement of robotic equipment. They also are responsible for safety rules and practices and for ensuring that robotic equipment is used properly.

CAD/CAM engineers (computer-aided design/computer-aided manufacturing) are experts in automated production processes. They design and supervise manufacturing systems that utilize robots and other automated equipment.

Manufacturing engineers manage the entire production process. They may evaluate production operations to determine whether robots can be used in an assembly line and make recommendations on purchasing robotic equipment. Some manufacturing engineers design robots. Other engineers specialize in a specific area of robotics, such as artificial intelligence, vision systems, and sensor systems. These specialists are developing robots with "brains" that are similar to those of humans.

Robotics technicians assist in all phases of robotics engineering. They install, repair, and maintain finished robots. Others help design and develop new kinds of robotics equipment. Technicians who install, repair, and maintain robots and robotic equipment need knowledge of electronics, electrical circuitry, mechanics, pneumatics, hydraulics, and computer programming. They use hand and power tools, testing instruments, manuals, schematic diagrams, and blueprints.

Before installing new equipment, technicians review the work order and instructional information; verify that the intended site in the factory is correctly supplied with the necessary electrical wires, switches, circuit breakers, and other parts; position and secure the robot in place, sometimes using a crane or other large tools and equipment; and attach various cables and hoses, such as those that connect a hydraulic power unit with the robot. After making sure that the equipment is operational, technicians program the robot for specified tasks, using their knowledge of its programming language. They may write the detailed instructions that program robots or reprogram a robot when changes are needed.

Once robots are in place and functioning, they may develop problems. Technicians then test components and locate faulty parts. When the problem is found, they may replace or recalibrate parts. Sometimes they suggest changes in circuitry or programming, or may install different end-of-arm tools on robots to allow machines to perform new functions. They may train robotics operators in how to operate robots and related equipment and help establish in-house basic maintenance and repair programs at new installations.

Companies that only have a few robots don't always hire their own robotics technicians. Instead they use *robot field technicians* who work for a robotic manufacturer. These technicians travel to manufacturing sites and other locations where robots are used to repair and service robots and robotic equipment.

Technicians involved with the design and development of new robotic devices are sometimes referred to as *robotics design technicians.* As part of a design team, they work closely with robotics engineers. The robotics design job starts as the engineers analyze the tasks and settings to be assigned and decide what kind of robotics system will best serve the necessary functions. Technicians involved with robot assembly, sometimes referred to as *robot assemblers,* commonly specialize in one aspect of robot assembly. *Materials handling technicians* receive requests for components or materials, then locate and deliver them to the technicians doing the actual assembly or those performing tests on these materials or components. *Mechanical assembly technicians* put together components and subsystems and install them in the robot. *Electrical assembly technicians* do the same work as mechanical assembly technicians but specialize in electrical components such as circuit boards and automatic switching devices. Finally, some technicians test the finished assemblies to make sure the robot conforms to the original specifications.

Other kinds of robotics technicians include *robot operators,* who operate robots in specialized settings, and *robotics trainers,* who train other employees in the installation, use, and maintenance of robots.

Robotics technicians may also be referred to as *electromechanical technicians, manufacturing technicians, robot mechanics, robotics repairmen, robot service technicians,* and *installation robotics technicians.*

Requirements

High School

In high school, students should take as many science, math, and computer classes as possible. Recommended courses are biology, chemistry, physics, algebra, trigonometry, geometry, calculus, graphics, computer science, English, speech, composition, social studies, and drafting. In addition, students should take shop and vocational classes that teach blueprint and electrical schematic reading, the use of hand tools, drafting, and the basics of electricity and electronics.

Postsecondary Training

Because changes occur so rapidly within this field, it is often recommended that engineers and technicians get a broad-based education that encompasses robotics but does not focus solely on robotics. Programs that provide the widest career base are those in automated manufacturing, which includes robotics, electronics, and computer science.

In order to become an engineer it is necessary to earn a bachelor of science degree. These programs generally take four or five years to complete. More than 400 colleges and universities offer courses in robotics or related technology. Many different types of programs are available. Some colleges and universities offer robotics engineering degrees and others offer engineering degrees with concentrations or options in robotics and manufacturing engineering. For some higher-level jobs, such as robotics designer, a master of science or doctoral degree is required.

Although the minimum educational requirement for a robotics technician is a high school diploma, many employers prefer to hire technicians who have received formal training beyond high school. Two-year programs are available in community colleges and technical institutes that grant an associate's degree upon completion. The armed forces also offer technical programs that result in associate's degrees in electronics, biomedical equipment, and computer science. The military uses robotics and other advanced equipment and offers excellent training opportunities to members of the armed forces. This training is highly regarded by many employers and can be an advantage in obtaining a civilian job in robotics.

Other Requirements

Because the field of robotics is rapidly changing, one of the most important requirements for a person interested in a career in robotics is the willingness to pursue additional training on an ongoing basis during his or her career. After completing their formal education, engineers and technicians may need to take additional classes in a college or university or take advantage of training offered through their employers and professional associations.

People planning on becoming robotics technicians need manual dexterity, good hand-eye coordination, and mechanical and electrical aptitudes.

Exploring

People interested in robotics can explore this field in many different ways. Because it is such a new field, it is important to learn as much as possible about current trends and recent technologies. Reading books and articles in trade magazines provides an excellent way to learn about what is happening in robotics technologies and expected future trends. Trade magazines with informative articles include *Robotics Engineering, Robotics Quarterly, Personal Robotics Magazine,* and *Robotics Today.*

Students can become robot hobbyists and build their own robots or buy toy robots and experiment with them. Complete robot kits are available through a number of companies and range from simple, inexpensive robots to highly complex robots with advanced features and accessories. A number of books that give instructions and helpful hints on building robots can be found at most public libraries and bookstores. In addition, relatively inexpensive and simple toy robots are available from electronics shops, department stores, and mail order companies.

Students can also participate in competitions. The Robotics International of the Society of Manufacturing Engineers sponsors a contest called the Student Robotics Automation Contest. Held every year, it is open to middle school through university-level students. Eight different categories challenge students in areas such as problem-solving skills, robot construction, and teamwork ability. Another annual competition, the International Aerial Robotics Competition, is sponsored by the Association for Unmanned Vehicle Systems. This competition, which requires teams of students to build complex robots, is open to college students. (Addresses for both associations are listed at the end of this article.)

Another great way to learn about robotics is to attend trade shows. Many robotics and automated machinery manufacturers exhibit their products at shows and conventions. Numerous such trade shows are held every year in different parts of the country. Information about these trade shows is available through association trade magazines and periodicals such as *Managing Automation.*

Other activities that foster knowledge and skills relevant to a career in robotics include membership in high school science clubs, participation in science fairs, and pursuing hobbies that involve electronics, mechanical equipment, and model building.

Employers

Robotics engineers and technicians are employed in virtually every manufacturing industry. Within the trend toward automation continuing—often via the use of robots—people trained in robotics can expect to find employment with almost all types of manufacturing companies in the future.

Starting Out

Many people entered robotics technician positions in the 1980s and early 1990s who were formerly employed as automotive workers, machinists, millwrights, computer repair technicians, and computer operators. In the mid-1990s, some companies began retraining current employees to troubleshoot and repair robots rather than hiring new workers. Because of these trends, entry-level applicants without any work experience may have difficulty finding their first jobs. Students who have participated in a cooperative work program or internship have the advantage of some work experience.

Graduates of two- and four-year programs may learn about available openings through their schools' job placement services. It also may be possible to learn about job openings through want ads in newspapers and trade magazines.

In many cases, it will be necessary to research companies that manufacture or use robots and apply directly to them. A number of directories are available that list such companies. One such directory is *Robotics and Vision Supplier Directory*. It is available for purchase from the Robotic Industries Association. Other directories that may be available at public or college libraries include *The CAD/CAM Industry Directory* and *Robotics, CAD/CAM Marketplace*.

Job opportunities may be good at small start-up companies or a start-up robotics unit of a large company. Many times these employers are willing to hire inexperienced workers as apprentices or assistants. Then, when their sales and production grow, these workers have the best chances for advancement.

Robotics engineers also may have difficulty finding their first jobs as many companies have retrained existing engineers in robotics. Job hunters may learn about job openings through their colleges' job placement services, advertisements in professional magazines and newspapers, or job fairs. In addition, recruiters may come to colleges to interview graduating students for prospective positions. In many cases, though, applicants will need to research a company using robotics engineers and apply directly to it.

Advancement

Engineers may start as part of an engineering team and do relatively simple tasks under the supervision of a project manager or more experienced engineer. With experience and demonstrated competency, they can move into higher engineering positions. Engineers who demonstrate good interpersonal skills, leadership abilities, and technical expertise may become team leaders, project managers, or chief engineers. Engineers can also move into supervisory or management positions. Some engineers pursue an MBA (master of business administration) degree. These engineers are able to move into top management positions. Some engineers also develop specialties, such as artificial intelligence, and move into highly specialized engineering positions.

After several years on the job, robotics technicians who have demonstrated their ability to handle more responsibility may be assigned some supervisory work or, more likely, will train new technicians. Experienced technicians and engineers may teach courses at their workplace or find teaching opportunities at a local school or community college.

Other routes for advancement include becoming a sales representative for a robotics manufacturing or design firm or working as an independent contractor for companies that use or manufacture robots.

With additional training and education, such as a bachelor's degree, technicians can become eligible for positions as robotics engineers.

Earnings

Earnings and benefits in manufacturing companies vary widely based on the size of the company, geographic location, nature of the production process, and complexity of the robots. In general, engineers with a bachelor of science degree earn annual salaries between $32,000 and $35,000 in their first job after graduation. Engineers with several years of experience earn salaries ranging from $35,000 to $60,000 a year.

Robotics technicians who are graduates of a two-year technical program earn between $22,000 and $26,000 a year. With increased training and experience, technicians can earn much more. Technicians with special skills, extensive experience, or added responsibilities can earn $36,000 or more. Technicians involved in design and training generally earn the highest salaries, with experienced workers earning $45,000 or more a year; those involved with maintenance and repair earn relatively less, with some beginning at salaries around $20,000 a year.

Employers offer a variety of benefits that can include the following: paid holidays, vacations, personal days, and sick leave; medical, dental, disability, and life insurance; 401-K plans, pension and retirement plans; profit sharing; and educational assistance programs.

Work Environment

Robotics engineers and technicians may work either for a company that manufactures robots or a company that uses robots. Most companies that manufacture robots are relatively clean, quiet, and comfortable environments. Engineers and technicians may work in an office or on the production floor. A large number of robotics manufacturers are found in California, Michigan, Illinois, Indiana, Pennsylvania, Ohio, Connecticut, Texas, British Columbia, and Ontario, although companies exist in many other states and parts of Canada.

Engineers and technicians who work in a company that uses robots may work in noisy, hot, and dirty surroundings. Conditions vary based on the type of industry within which one works. Automobile manufacturers use a significant number of robots, as well as manufacturers of electronics components and consumer goods and the metalworking industry. Workers in a foundry work around heavy equipment and in hot and dirty environments. Workers in the electronics industry generally work in very clean and quiet environments. Some robotics personnel are required to work in clean room environments, which keep electronic components free of dirt and other contaminants. Workers in these environments wear face masks, hair coverings, and special protective clothing.

Some engineers and technicians may confront potentially hazardous conditions in the workplace. Robots, after all, are often designed and used precisely because the task they perform involves some risk to humans: handling laser beams, arc-welding equipment, radioactive substances, or hazardous chemicals. When they design, test, build, install, and repair robots, it is inevitable that some engineers and technicians will be exposed to these same risks. Plant safety procedures protect the attentive and cautious worker, but carelessness in such settings can be especially dangerous.

In general, most technicians and engineers work 40-hour workweeks, although overtime may be required for special projects or to repair equipment that is shutting down a production line. Some technicians, particularly those involved in maintenance and repairs, may work shifts that include evening, late night, or weekend work.

Field service technicians travel to manufacturing sites to repair robots. Their work may involve extensive travel and overnight stays. They may work at several sites in one day or stay at one location for an extended period for more difficult repairs.

Outlook

Employment opportunities for robotics engineers and technicians are closely tied to economic conditions in the United States and in the global marketplace. During the late 1980s and early 1990s, the robotics market suffered because of a lack of orders for robots and robotic equipment and intense foreign competition, especially from Japanese manufacturers.

In addition, many manufacturers reduced their number of employees significantly. Some companies cut their staffs by as many as 70,000 employees. Many companies that use large numbers of robotics personnel, such as those in the automobile industry, are represented by unions. These unions have agreements with companies that humans will not lose their jobs because of automation. When companies do downsize, existing workers are given priority and are retrained to work in positions such as robotics technicians. This makes it more difficult for inexperienced workers to enter the field.

However, in 1995, U.S. robot manufacturers shipped record numbers of robots and robotic equipment. In addition, some Japanese robot builders shifted production to U.S. facilities. After a slump of several years, it is expected that the robotics industry will once again pick up. Some robotics industry experts are predicting that during the next decade there will be a robotics boom and robotics sales will double.

The use of industrial robots is expected to grow as robots become more programmable and flexible and as manufacturing processes become more automated. Growth is also expected in nontraditional applications, such as education, health care, security, and nonindustrial purposes.

It is difficult to predict whether recent sales and the rising production of robots will increase employment opportunities, but trends for automated manufacturing equipment and a willingness by manufacturers to invest in capital expenditures are promising signs of growth. For prospective robotics engineers and technicians, this expansion suggests that more workers will be needed to design, build, install, maintain, repair, and operate robots.

For More Information

For information on educational programs and student membership in the International Service Robot Association and Global Automation Information Network, contact:

Robotic Industries Association
PO Box 3724
Ann Arbor, MI 48106
Tel: 313-994-6088
Web: http://www.robotics.org

For information on educational programs, competitions, and student membership in SME, Robotics International, or Machine Vision Association, contact:

Society of Manufacturing Engineers (SME)
Education Department
PO Box 930
Dearborn, MI 48121-0930
Tel: 313-271-1500
Web: http://www.sme.org

For information on careers and educational programs, contact:

Robotics and Automation Council
Institute of Electrical and Electronics Engineers (IEEE)
345 East 47th Street
New York, NY 10017
Tel: 212-705-7900
Web: http://www.ieee.org

For information on competitions and student membership, contact:

Association for Unmanned Vehicle Systems
1735 North Lynn Street, Suite 950
Arlington, VA 22209
Tel: 703-524-6646

Traffic Engineers

	School Subjects
Geography Mathematics	
	Personal Skills
Communication/ideas Technical/scientific	
	Work Environment
Indoors and outdoors Primarily multiple locations	
	Minimum Education Level
Bachelor's degree	
	Salary Range
$34,800 to $51,600 to $86,400	
	Certification or Licensing
None available	
	Outlook
Faster than the average	

Overview

Traffic engineers study factors that influence traffic conditions on roads and streets, including street lighting, visibility and location of signs and signals, entrances and exits, and the presence of factories or shopping malls. They use this information to design and implement plans and electronic systems that improve the flow of traffic.

History

During the early colonial days, dirt roads and Native American trails were the primary means of land travel. In 1806, the U.S. Congress provided for the construction of the first road, known as the Cumberland Road. More and more roads were built, connecting neighborhoods, towns, cities, and states. As the population increased and modes of travel began to advance, more roads were needed to facilitate commerce, tourism, and daily transportation. Electric traffic signals were introduced in the United States in 1928 to help control automobile traffic. Because land travel was becoming increasingly complex, traffic engineers were trained to ensure safe travel on roads and highways, in detours and construction work zones, and for special events such as sports competitions and presidential conventions, among others.

The Job

Traffic engineers study factors such as signal timing, traffic flow, high-accident zones, lighting, road capacity, and entrances and exits in order to increase traffic safety and to improve the flow of traffic. In planning and creating their designs, engineers may observe such general traffic influences as the proximity of shopping malls, railroads, airports, or factories, and other factors that affect how well traffic moves. They apply standardized mathematical formulas to certain measurements to compute traffic signal duration and speed limits, and they prepare drawings showing the location of new signals or other traffic control devices. They may perform statistical studies of traffic conditions, flow, and volume, and may—on the basis of such studies—recommend changes in traffic controls and regulations. Traffic engineers design improvement plans with the use of computers and through on-site investigation.

Traffic engineers address a variety of problems in their daily work. They may conduct studies and implement plans to reduce the number of accidents on a particularly dangerous section of highway. They might be asked to prepare traffic impact studies for new residential or industrial developments, implementing improvements to manage the increased flow of traffic. To do this, they may analyze and adjust the timing of traffic signals, suggest the widening of lanes, or recommend the introduction of bus or carpool lanes. In the performance of their duties, traffic engineers must be constantly aware of the effect their designs will have on nearby pedestrian traffic and on environmental concerns, such as air quality, noise pollution, and the presence of wetlands and other protected areas.

Traffic engineers use computers to monitor traffic flow onto highways and at intersections, to study frequent accident sites, to determine road and highway capacities, and to control and regulate the operation of traffic signals throughout entire cities. Computers allow traffic engineers to experiment with multiple design plans while monitoring cost, impact, and efficiency of a particular project.

Traffic engineers who work in government often design or oversee roads or entire public transportation systems. They might oversee the design, planning, and construction of new roads and highways or manage a system that controls the traffic signals by the use of a computer. Engineers frequently interact with a wide variety of people, from average citizens to business leaders and elected officials.

Traffic technicians assist traffic engineers. They collect data in the field by interviewing motorists at intersections where traffic is often congested or where an unusual number of accidents have occurred. They also use radar equipment or timing devices to determine the speed of passing vehicles at certain locations, and they use stopwatches to time traffic signals and other delays to traffic. Some traffic technicians may also have limited design duties.

Requirements

High School

High school students interested in a traffic engineering career should have mathematical skills through training in algebra, logic, and geometry and a good working knowledge of statistics. They should have language skills that will enable them to write extensive reports making use of statistical data, and they should be able to present such reports before groups of people. They should also be familiar with computers and electronics in general. Traffic engineers should have a basic understanding of the workings of government since they must frequently address regulations and zoning laws and meet and work with government officials. A high school diploma is the minimum educational requirement for traffic technicians.

Postsecondary Training

Traffic engineers must have a bachelor's degree in civil, electrical, mechanical, or chemical engineering. Because the field of transportation is so vast, many engineers have educational backgrounds in science, planning, computers, environmental planning, and other related fields. Educational courses for traffic engineers in transportation may include transportation planning, traffic engineering, highway design, and related courses such as computer science, urban planning, statistics, geography, business management, public administration, and economics.

Traffic engineers acquire some of their skills through on-the-job experience and training conferences and mini-courses offered by their employers, educational facilities, and professional engineering societies. Traffic technicians receive much of their training on the job and through education courses offered by various engineering organizations.

Certification or Licensing

Currently, no certification exists in the field of traffic engineering. The Institute of Transportation Engineers (ITE) is working on a certification program, which it hopes to implement in the near future.

Other Requirements

Traffic engineers should enjoy the challenge of solving problems. They should have good oral and written communication skills since they frequently work with others. Engineers must also be creative and able to visualize the future workings of their designs; that is, how they will improve traffic flow, effects on the environment, and potential problems.

Exploring

Interested students can join a student chapter of the ITE to see if a career in transportation engineering is for them. An application for student membership in the ITE can be obtained by writing the association at the address listed in the "For More Information" section at the end of this article.

Employers

Traffic engineers are employed by federal, state, or local agencies or as private consultants by states, counties, towns, and even neighborhood groups. Many teach or engage in research in colleges and universities.

Starting Out

The ITE offers a resume service to students that are members of the organization. Student members can get their resumes published in the *ITE Journal*. The journal also lists available positions for traffic engineering positions throughout the country. Most colleges also offer job placement programs to help traffic engineering graduates locate their first jobs.

Advancement

Experienced traffic engineers may advance to become directors of transportation departments or directors of public works in civil service positions. A vast array of related employment in the transportation field is available for those engineers who pursue advanced or continuing education. Traffic engineers may specialize in transportation planning, public transportation (urban and intercity transit), airport engineering, highway engineering, harbor and port engineering, railway engineering, or urban and regional planning.

Earnings

Salaries for traffic engineers vary widely depending upon duties, qualifications, and experience. According to a salary survey by the ITE, professional entry-level junior traffic engineers (Level I) earn starting annual salaries of $34,772. Level II traffic engineers, with a minimum of two year's experience and who oversaw small projects, earned $41,318 per year. Level III engineers, who supervise others and organize small to mid-size projects, earn annual salaries of $51,563. Level IV engineers, who are responsible for the supervision of large projects, staffing, and scheduling, earn annual salaries of $61,908. Those traffic engineers who have titles such as director of traffic engineering, director of transportation planning, professor, or vice president (Level V) earn average salaries of $72,867 per year. Level VI engineers who have advanced to upper-level management positions, such as president, general manager, director of transportation or public workers, and who are responsible for major decision-making, earn the highest salaries: $86,375 per year. Traffic engineers are also eligible for paid vacation, sick, and personal days, health insurance, pension plans, and in some instances, profit sharing.

Work Environment

Traffic engineers perform their duties both indoors and outdoors, under a variety of conditions. They are subject to the noise of heavy traffic and various weather conditions while gathering data for some of their studies. They may speak to a wide variety of people as they check the success of their designs. Traffic engineers also spend a fair amount of time in the quiet of an office, making calculations and analyzing the data they have collected in the field. They also spend a considerable amount of time working with computers to optimize traffic signal timing, in general design, and to predict traffic flow.

Traffic engineers must be comfortable working with other professionals, such as traffic technicians, designers, planners, and developers, as they work to create a successful transportation system. At the completion of a project they can take pride in the knowledge that they have made the streets, roads, and highways safer and more efficient as a result of their designs.

Outlook

There were nearly 24,000 traffic engineers in the United States in the 1990s. Employment for traffic engineers is expected to increase faster than the average through 2006. More engineers will be needed to work with ITS (Intelligent Transportation System) technology such as electronic toll collection, cameras for traffic incidents/detection, and fiber optics for use in variable message signs. As the population increases and continues to move to suburban areas, qualified traffic engineers will be needed to analyze, assess, and implement traffic plans and designs to ensure safety and the steady, continuous flow of traffic. In cities, traffic engineers will continue to be needed to staff advanced transportation management centers that oversee vast stretches of road using computers, sensors, cameras, and other electrical devices.

For More Information

For information regarding fellowships, seminars, tours, and general information concerning the transportation engineering field, contact:

American Association of State Highway and Transportation Officials
444 North Capitol Street, NW
Washington, DC 20001
Tel: 202-624-5800

American Public Transportation Association
1201 New York Avenue, NW, Suite 400
Washington, DC 20005
Tel: 202-898-4000

Institute of Transportation Engineers
525 School Street, SW, Suite 410
Washington, DC 20024-2729
Tel: 202-554-8050
Web: http://www.ite.org

U.S. Department of Transportation
400 Seventh Street, SW
Washington, DC 20590
Tel: 202-366-4000

Index